Mathematical Analysis of Viscoelastic Flows

CBMS-NSF REGIONAL CONFERENCE SERIES IN APPLIED MATHEMATICS

A series of lectures on topics of current research interest in applied mathematics under the direction of the Conference Board of the Mathematical Sciences, supported by the National Science Foundation and published by SIAM.

GARRETT BIRKHOFF, *The Numerical Solution of Elliptic Equations*
D. V. LINDLEY, *Bayesian Statistics, A Review*
R. S. VARGA, *Functional Analysis and Approximation Theory in Numerical Analysis*
R. R. BAHADUR, *Some Limit Theorems in Statistics*
PATRICK BILLINGSLEY, *Weak Convergence of Measures: Applications in Probability*
J. L. LIONS, *Some Aspects of the Optimal Control of Distributed Parameter Systems*
ROGER PENROSE, *Techniques of Differential Topology in Relativity*
HERMAN CHERNOFF, *Sequential Analysis and Optimal Design*
J. DURBIN, *Distribution Theory for Tests Based on the Sample Distribution Function*
SOL I. RUBINOW, *Mathematical Problems in the Biological Sciences*
P. D. LAX, *Hyperbolic Systems of Conservation Laws and the Mathematical Theory of Shock Waves*
I. J. SCHOENBERG, *Cardinal Spline Interpolation*
IVAN SINGER, *The Theory of Best Approximation and Functional Analysis*
WERNER C. RHEINBOLDT, *Methods of Solving Systems of Nonlinear Equations*
HANS F. WEINBERGER, *Variational Methods for Eigenvalue Approximation*
R. TYRRELL ROCKAFELLAR, *Conjugate Duality and Optimization*
SIR JAMES LIGHTHILL, *Mathematical Biofluiddynamics*
GERARD SALTON, *Theory of Indexing*
CATHLEEN S. MORAWETZ, *Notes on Time Decay and Scattering for Some Hyperbolic Problems*
F. HOPPENSTEADT, *Mathematical Theories of Populations: Demographics, Genetics and Epidemics*
RICHARD ASKEY, *Orthogonal Polynomials and Special Functions*
L. E. PAYNE, *Improperly Posed Problems in Partial Differential Equations*
S. ROSEN, *Lectures on the Measurement and Evaluation of the Performance of Computing Systems*
HERBERT B. KELLER, *Numerical Solution of Two Point Boundary Value Problems*
J. P. LASALLE, *The Stability of Dynamical Systems* - Z. ARTSTEIN, *Appendix A: Limiting Equations and Stability of Nonautonomous Ordinary Differential Equations*
D. GOTTLIEB AND S. A. ORSZAG, *Numerical Analysis of Spectral Methods: Theory and Applications*
PETER J. HUBER, *Robust Statistical Procedures*
HERBERT SOLOMON, *Geometric Probability*
FRED S. ROBERTS, *Graph Theory and Its Applications to Problems of Society*
JURIS HARTMANIS, *Feasible Computations and Provable Complexity Properties*
ZOHAR MANNA, *Lectures on the Logic of Computer Programming*
ELLIS L. JOHNSON, *Integer Programming: Facets, Subadditivity, and Duality for Group and Semi-Group Problems*
SHMUEL WINOGRAD, *Arithmetic Complexity of Computations*
J. F. C. KINGMAN, *Mathematics of Genetic Diversity*
MORTON E. GURTIN, *Topics in Finite Elasticity*
THOMAS G. KURTZ, *Approximation of Population Processes*

Michael Renardy
Virginia Polytechnic Institute and State University
Blacksburg, Virginia

Mathematical Analysis of Viscoelastic Flows

SOCIETY FOR INDUSTRIAL AND APPLIED MATHEMATICS
PHILADELPHIA

10 9 8 7 6 5 4 3 2 1

Library of Congress Cataloging-in-Publication Data

Renardy, Michael.
Mathematical analysis of viscoelastic flows / Michael Renardy.
p. cm. — (CBMS-NSF regional conference series in applied mathematics ; 73)
Includes bibliographical references and index.
ISBN 0-89871-457-5 (pbk.)
1. Viscous flow. 2. Viscoelasticity. I. Title. II. Series.

QA929 .R43 2000
532′.0533—dc21

00-036558

Contents

Preface

Fluids with complex microstructure, such as polymers, suspensions, and granular materials, abound in daily life and in many industrial processes, e.g., in the chemical, food, and oil industries. The phenomena and mathematical models for such fluids are much more varied and complex than those of traditional Newtonian fluid dynamics. This monograph presents an overview of mathematical problems, methods, and results arising in the study of flows of polymeric liquids, such as molten plastics, lubricants, paints, and many biological fluids. The role of numerical simulation in the study of such flows has increased tremendously over the past 15 years, and the phenomena and numerical difficulties in complex flows, e.g., instabilities, boundary layers, and singularities, have led to new and challenging mathematical questions.

The monograph begins with an introduction to phenomena observed in viscoelastic flows (Chapter 1), the formulation of mathematical equations to model such flows (Chapter 2), and the behavior of various models in simple flows (Chapter 3). We then review results on basic existence questions (Chapter 4). Chapter 5 provides a brief introduction to some of the issues arising in numerical simulation. Chapter 6 discusses the asymptotics of the high Weissenberg limit, and Chapter 7 uses these asymptotic ideas to study the behavior near reentrant corners, which arise, for instance, in contraction flows. Chapter 8 discusses the analysis of flow instabilities. In Chapter 9, we discuss change of type in the equations of viscoelastic flow and applications to melt fracture and delayed die swell. Chapter 10 is concerned with jets and filaments and their breakup.

The monograph is based on a series of lectures presented at the CBMS-NSF conference on Mathematical Analysis of Viscoelastic Flows held at the University of Delaware, June 19–23, 1999. I thank the National Science Foundation for support of this conference and for support of much of the research reviewed in this monograph. Thanks are also due to the organizers of the conference, David Olagunju and Yuriko Renardy, and to Pamela Cook and Gilberto Schleiniger, who made the local arrangements.

Michael Renardy

Chapter 1

Phenomena in Non-Newtonian Flows

A fluid that's macromolecular
Is really quite weird – in particular
The abnormal stresses
The fluid possesses
Give rise to effects quite spectacular.
Robert Byron Bird [5]

Traditional Newtonian fluid mechanics is governed by the interplay of viscous and inertial forces and, where free surfaces are involved, surface tension. Daily life abounds with examples of fluids, however, whose complex microstructure produces a rich variety of behavior which falls outside the scope of Newtonian fluid mechanics. Often such fluids exhibit interesting and unexpected flow patterns and complex dynamics even in situations where inertia and surface tension play no role at all.

In essence, Newtonian fluids are fluids whose constituent particles are too small for their dynamics to interact substantially with the macroscopic motion. The Newtonian model becomes inadequate for fluids which have a microstructure involving much larger scales than the atomic scale. There are many possibilities for such microstructures: for instance, suspensions such as concrete and dough, foams, and liquid crystals; granular media such as sand, gravel, and coal; and flows of materials we ordinarily regard as solid, e.g., flowing glaciers. By far the most widely studied and best understood class of non-Newtonian fluids is that of polymeric fluids, and it is this class of fluids on which these lectures will concentrate. Examples include molten plastics, engine oils with polymeric additives, paints, and many biological fluids, e.g., egg white and blood.

The basic feature of polymeric fluids is the presence of long chain molecules. In a flow, these chain molecules are stretched out by the drag forces exerted on them by the surrounding fluid. The natural tendency of the molecule to retract from this stretched configuration generates an elastic force which contributes to the macroscopic stress tensor. In the second chapter, we shall discuss specifically how this interaction between flow and molecular dynamics can be modeled mathematically. In this chapter, we describe a number of phenomena in polymeric flows and their qualitative explanation. Most of the pictures of experiments shown in this chapter, and many more, can be found in [10].

1.1 Effects due to normal stresses

If a Newtonian fluid is subjected to shearing, it develops a friction force to which we refer as viscosity. In polymeric fluids, there is, in addition, an alignment of polymer molecules with the flow direction, which causes a tension force in the flow direction. This tension is referred to as a "normal stress." For a detailed mathematical description, I refer to the third chapter.

One of the most striking manifestations of normal stresses is the rod-climbing or Weissenberg effect. In the experiment, a rotating rod is inserted into a pool of liquid. In a Newtonian fluid, the rotating motion generates a centrifugal force pushing the liquid outward, and the free surface forms a dip near the rod. In polymeric fluids, in contrast, the normal stresses cause a tension along the concentric streamlines, which leads to a force pushing the fluid inward. Consequently, the free surface rises and the fluid climbs up the rod. See Figure 1.1 [10] for a picture of a typical experiment. Actually, all of us have done this experiment in our own kitchens.

Another effect linked to normal stresses is die swell. When a fluid exits from a pipe (referred to as a "die" in the language of polymer processing), the diameter of the free jet forming outside the pipe is not equal to the diameter of the pipe. In Newtonian fluids, the change depends on the Reynolds number. At low Reynolds numbers, there is a slight increase in diameter, but when inertial effects become strong, the diameter shrinks. Polymeric fluids tend to show a strong increase in diameter. See Figure 1.2 [10] for a contrast between a Newtonian and a polymeric die swell. The reason for this effect is again the tension along streamlines which is generated by the shearing motion inside the pipe. When the fluid exits, this tension is relieved, causing the jet to shrink in the longitudinal direction and expand in the transverse direction. In situations where both inertia and elasticity are important, die swell can be delayed, i.e., it happens not at the exit of the die, but farther down the jet; see Figure 1.3 [10].

1.2 Effects due to elongational viscosity

If a polymeric fluid is subjected to stretching, the polymer molecules become extended, leading to a large elastic force. A striking manifestation of this is the

Figure 1.1: The Weissenberg effect (Figure 2.1 of [10]). Originally published in H.A. Barnes, J.F. Hutton, and K.A. Walters, *An Introduction to Rheology*, Elsevier Science, 1989, Figure 4.7. Reprinted with permission from Elsevier Science.

tubeless siphon effect illustrated in Figure 1.4 [10]. Here, the force generated by the stretching in the falling jet is strong enough to pull the rest of the liquid out of the beaker—against gravity.

If the fluid was Newtonian, the jet would not only fail to pull the rest of the fluid along, it would quickly break. In contrast, jets or filaments of polymeric liquids are remarkably stable. We can observe this every time we eat pizza, and the stability of liquid filaments is also crucial for the feasibility of the manufacture of synthetic fibers. The reason for the stability of filaments is the strong resistance of polymers to elongation. To break a filament, it must be stretched locally to infinity. While this is easy to do for a Newtonian fluid, the elastic force of the polymers prevents it from happening, or at least delays it. The surface tension

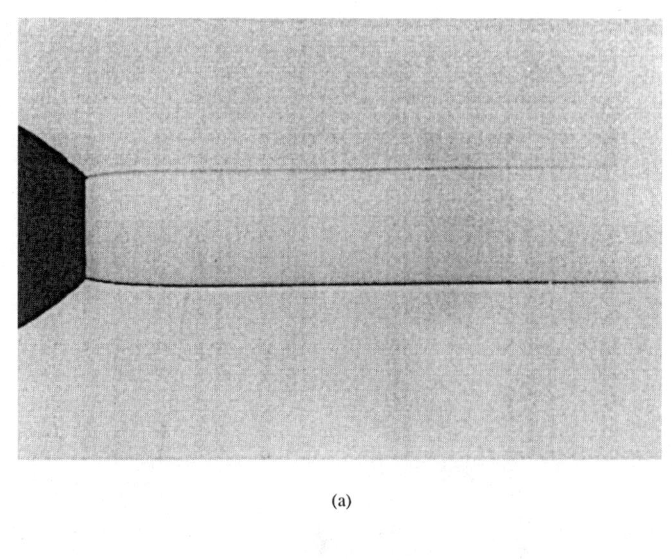

(a)

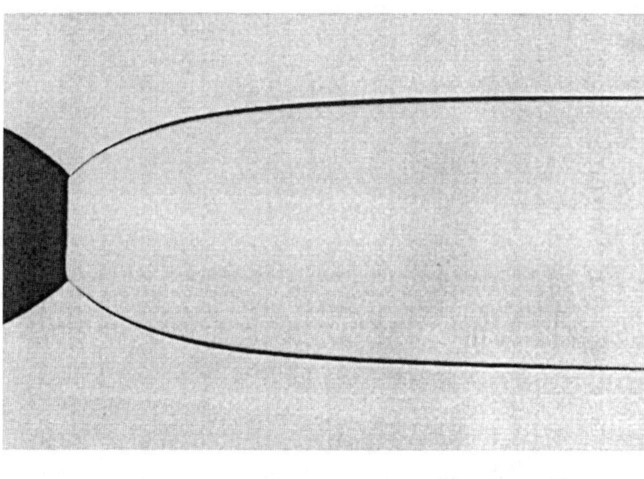

(b)

Figure 1.2: Die swell in Newtonian and polymeric liquids (Figure 2.10 of [10]). Reprinted with permission from Elsevier Science.

driven breakup of a jet illustrates this (Figure 1.5 [10]). While the Newtonian jet breaks into droplets, the polymeric jet evolves into a beads-on-a-string shape, where drops remain connected by thin filaments which persist for an extended time. We shall return to this problem in the tenth chapter. The stabilizing effect of polymers is used in controlling atomization of liquids by adding additives. The control of drop size is important, for instance, in improving the accuracy when fluids are sprayed at a target from the air, as in crop dusting or fire fighting.

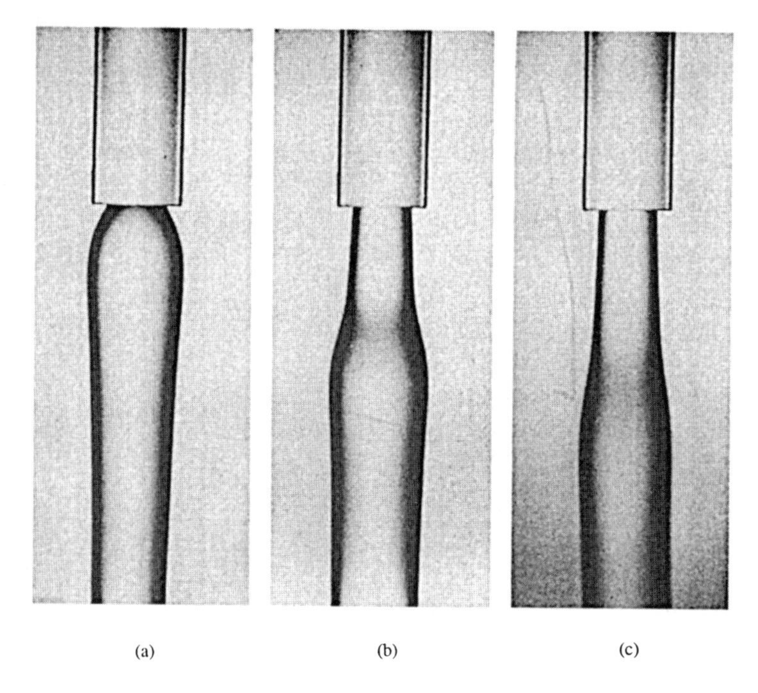

(a) (b) (c)

Figure 1.3: Delayed die swell (Figure 2.11 of [10]). Originally published in H. Giesekus, *Rheol. Acta* **8** (1968), 411. Reprinted with permission from Springer-Verlag.

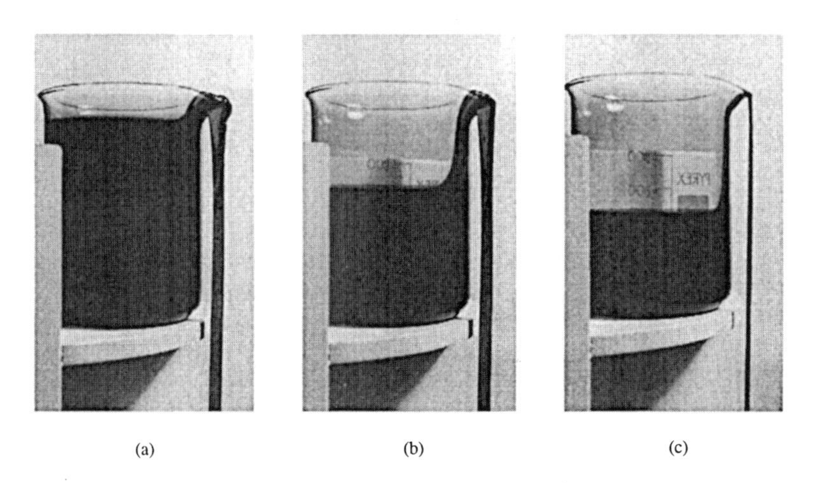

(a) (b) (c)

Figure 1.4: Tubeless siphon effect (Figure 2.17 of [10]). Reprinted with permission from Elsevier Science.

The phenomenon of drag reduction is another striking instance where polymers have a dramatic impact on flow behavior. If water carries even a small quantity of dissolved polymers, then the pressure required to pump a given

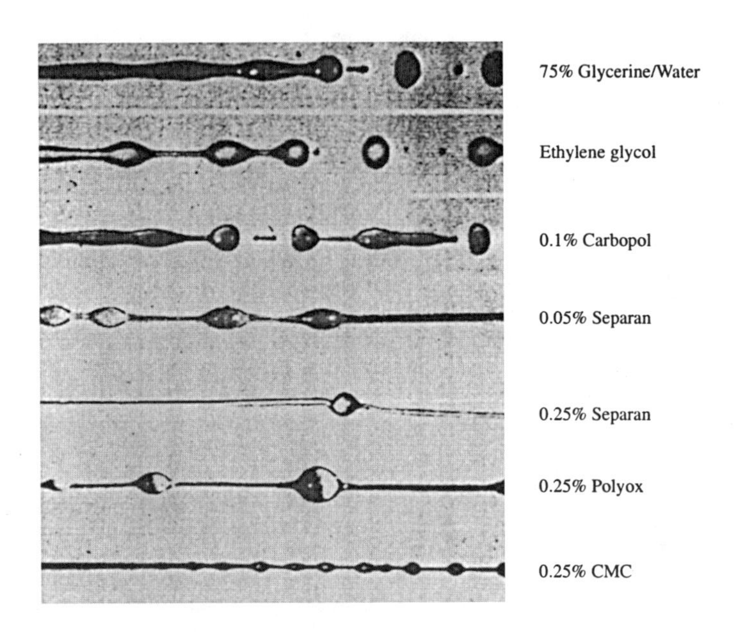

Figure 1.5: Surface tension driven jet breakup (Figure 6.1 of [10]). Originally published in M. Golding et al., Breakup of a laminar capillary jet of a viscoelastic fluid, *J. Fluid Mech.* **38** (1969), 689–711. Reprinted with permission from Cambridge University Press.

amount of water through a pipe can be reduced significantly. Evidently, the polymer suppresses the turbulence in the water to some extent, bringing the flow closer to the "optimal" laminar flow. With some polymers, the effect has been observed at concentrations as low as one part per million! Although the phenomenon has been known for the past 50 years, it has not yet been fully explained. The prevalent thinking among rheologists is that the elongational resistance of the polymer suppresses the vortices which produce turbulence, although there is considerable uncertainty about the strength of this effect under the conditions of drag reduction, and other hypotheses have been offered. Numerical simulations of drag-reducing flows have only recently become feasible; they appear to support the hypothesis that elongational viscosity is the determining factor [16].

1.3 Contraction flows

The transition from a wider pipe to a narrower one is typical of a situation that is frequently encountered in processing. A Stokes flow for this problem looks like the first picture in Figure 1.6 [10]; most of the fluid goes into the opening of the narrower pipe, but there is a recirculating vortex in the corner. Inertia has the effect of shrinking this vortex, while fluid elasticity significantly increases its size, as shown in the sequence of pictures in Figure 1.6. We can get a rough

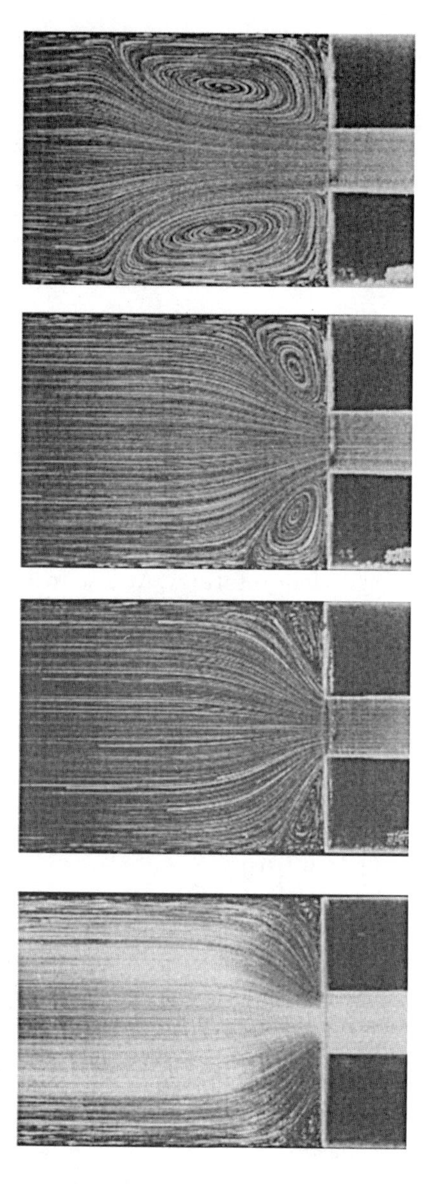

Figure 1.6: Contraction flows with varying levels of elasticity (Figure 3.2 of [10]). Originally published in D.V. Boger, D.U. Hur, and R.J. Binnington, *J. Non-Newt. Fluid Mech.* **20** (1986), 31. Reprinted with permission from Elsevier Science.

understanding of this if we consider the fact that the fluid has to speed up in order to enter the narrower pipe. In inertia-dominated flows, this speeding up of the fluid leads to a Bernoulli suction, which pulls fluid out of the corner and into the pipe, hence the decrease in the vortex. For elastic fluids, on the other hand, the speeding up along streamlines in the center causes an elastic tension along these streamlines, which exerts a pull on the fluid directly upstream and pushes the fluid on the sides back into the corner.

In many polymeric fluids, more complicated flows arise as the flow rate is increased. There may be flow instabilities leading to oscillating vortices and/or the formation of another vortex attached to the lip of the contraction. Contraction flows are a popular and challenging problem for numerical simulation. One of the mathematical issues in contraction flows is the nature of corner singularities; we shall return to this topic in the seventh chapter.

1.4 Instabilities

Flow instabilities and the patterns emerging from them are a rich subject in Newtonian fluid mechanics. Viscoelastic flows lead to new mechanisms of instability. Rotating shear flows have attracted much recent attention. One of the classical problems of hydrodynamic instability in the Newtonian case is the Taylor problem, which is the flow between concentric cylinders with the inner cylinder or possibly both cylinders rotating. At low rotation rates, the flow is along concentric circles, but at higher rotation rates, the centrifugal force causes an instability leading to Taylor cells (see Figure 1.7 [10]). In viscoelastic flows, there are patterns analogous to Taylor cells, but they are caused by normal stresses rather than centrifugal forces. Similar effects are found in other rotating shear flows, such as the cone-and-plate flow sketched in Figure 1.8 [10], which forms a basis for rheometers.

Normal stress differences between two fluids can lead to interfacial instabilities in two-layer shear flows. A deformed interface resulting from such an instability is shown in Figure 1.9 [104].

Figure 1.7: Taylor cells in Newtonian and viscoelastic flows (Figure 5.11 of [10]). Originally published in G.S. Beavers and D.D. Joseph, *Phys. Fluids* **17** (1974), 650. Reprinted with permission from the American Institute of Physics and from the authors.

Figure 1.8: Instability in cone-and-plate flow (Figure 5.8 of [10]). Originally published in D.F. Griffiths and K. Walters, On edge effects in rheometry, *J. Fluid Mech.* **42** (1970), 379–399. Reprinted with permission from Cambridge University Press.

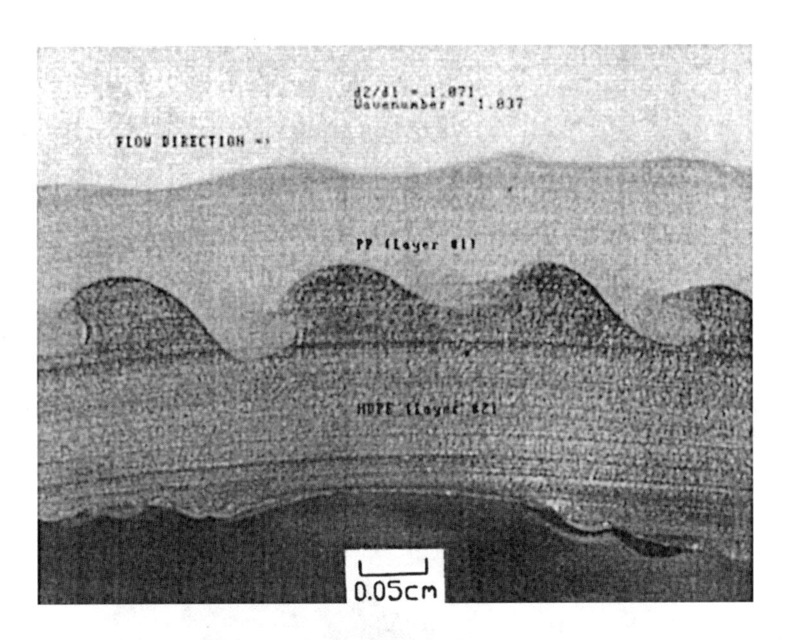

Figure 1.9: Interfacial instability driven by normal stress jump. Reprinted from [104] with permission from the Society of Rheology and from the authors.

The extrusion of polymer melts from an orifice often leads to instabilities when a critical flow rate is exceeded. The avoidance of such instabilities is of major commercial importance since the instability leads to unacceptable products in processing. There are fine-scale surface irregularities known as sharkskin, as well as gross distortions, which are called melt fracture. The onset of instability is generally coincident with a sudden increase in flow rate (spurt). Examples of these phenomena are shown in Figure 1.10 [10]. Although melt fracture has been studied for 40 years, a full theoretical explanation remains elusive. Most rheologists believe that the phenomenon is linked to wall slip, but the precise laws governing wall slip and the linkage between slip and instability are still very speculative. Another hypothesis that has been advanced is that of a constitutive instability, where there is a maximum in the shear stress versus shear rate curve and the material spurts when this maximum is reached.

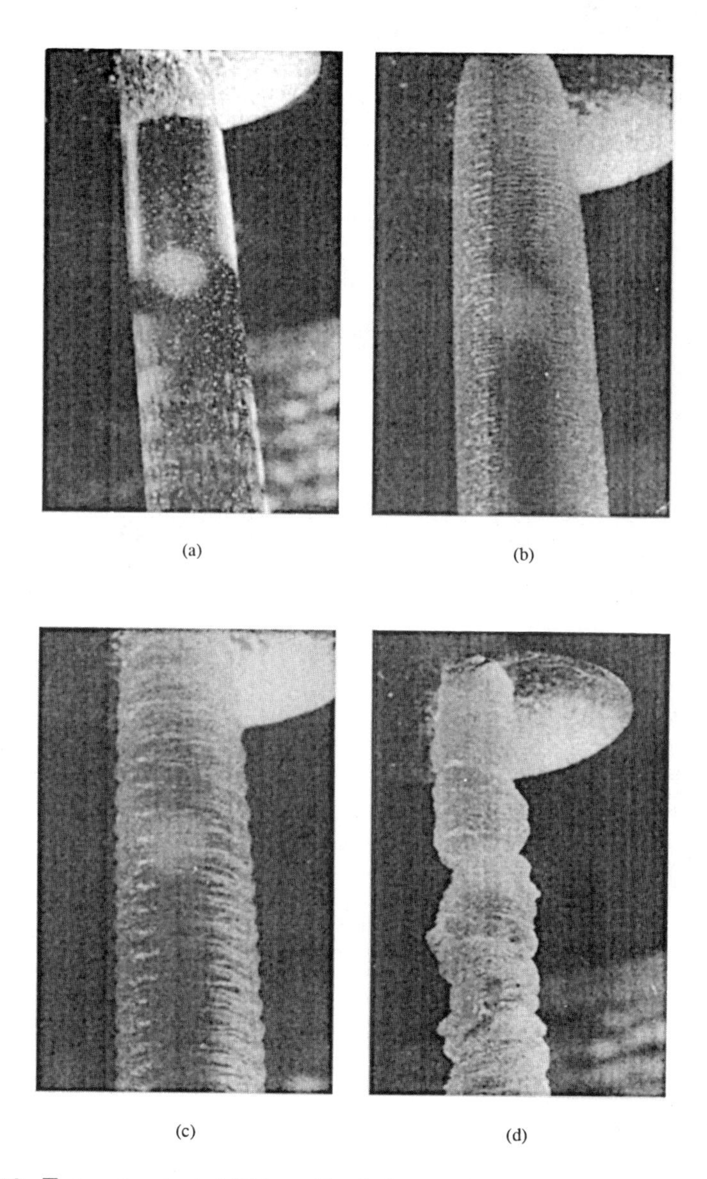

Figure 1.10: Extrusion instabilities: Sharkskin and melt fracture (Figure 2.16 of [10]). Originally published in J.M. Piau, N. El Kissi, and B. Tremblay, *J. Non-Newt. Fluid Mech.* **34** (1990), 145. Reprinted with permission from Elsevier Science.

Chapter 2

Mathematical Formulation

2.1 Balance laws

The motion of every fluid is governed by the conservation of mass and momentum, and, if thermal effects are important, the balance of energy. In these lectures, we shall concern ourselves mostly with purely mechanical problems where we assume constant temperature. In many processing operations, on the other hand, temperature variations are considerable, and the dependence of rheological parameters on temperature can be a very large factor. We shall also assume that the fluids are incompressible. Compressibility is seldom of importance in flows of polymeric liquids. Hence the conservation of mass assumes the simple form

$$\operatorname{div} \mathbf{v} = 0, \tag{2.1}$$

where $\mathbf{v}$ is the velocity of the fluid.

In addition, we have the balance of momentum

$$\rho\Big(\frac{\partial \mathbf{v}}{\partial t} + (\mathbf{v} \cdot \nabla)\mathbf{v}\Big) = \operatorname{div} \mathbf{T} - \nabla p. \tag{2.2}$$

Here ρ is the density, $\mathbf{T}$ is the extra stress tensor, and p is an isotropic pressure. The left side of the equation represents the force of inertia, i.e., the density times the acceleration of a fluid particle. While this term is responsible for most of the excitement in Newtonian fluid mechanics, it is often negligible in polymeric fluids. The stress tensor $\mathbf{T}$ represents the forces which the material develops in response to being deformed. To complete the mathematical formulation, we need a constitutive law relating $\mathbf{T}$ to the motion (in principle, $\mathbf{T}$ can also depend on the pressure; this is sometimes important in polymer melt processing where very high pressures may be reached).

2.2 Linear models

The constitutive law for a Newtonian fluid is given by

$$\mathbf{T} = \eta(\nabla \mathbf{v} + (\nabla \mathbf{v})^T). \tag{2.3}$$

Here η is a constant known as the viscosity. In viscoelastic fluids, the stresses depend not only on the current motion of the fluid, but on the history of the motion. If we assume that this dependence is linear, it is natural to express it in integral form:

$$\mathbf{T}(\mathbf{x},t) = \int_{-\infty}^{t} G(t-s)(\nabla \mathbf{v}(\mathbf{x},s) + (\nabla \mathbf{v}(\mathbf{x},s))^T)\, ds. \tag{2.4}$$

This is Boltzmann's [11] theory of linear viscoelasticity. The function G is called the stress relaxation modulus; its derivative $m = -G'$ is referred to as the memory function. It is usually assumed that G is positive and monotone decreasing; indeed, all examples ever considered seriously by rheologists are completely monotone functions. If we want to recover the Newtonian fluid as a special case, we must also allow a singular contribution to G that is a multiple of the delta function. In general, the viscosity of a linear viscoelastic fluid is the integral of G:

$$\eta = \int_0^\infty G(s)\, ds. \tag{2.5}$$

Maxwell's theory of linear viscoelasticity [56] assumes instead that the stress is linked to the velocity gradient by an ordinary differential equation

$$\mathbf{T}_t + \lambda \mathbf{T} = \mu(\nabla \mathbf{v} + (\nabla \mathbf{v})^T). \tag{2.6}$$

We can solve the differential equation and express the result in the form (2.4); the stress relaxation modulus is $G(s) = \mu \exp(-\lambda s)$. We can also formulate "multimode" Maxwell models, where the stress is a linear superposition of several terms that are each determined by an equation of the form (2.6) (or, equivalently, the stress relaxation modulus is a linear combination of exponentials), and "Jeffreys models," for which the stress is a linear combination of a Maxwell and a Newtonian term.

The quantity $1/\lambda$ in (2.6) has the dimension of time and is known as a relaxation time. It is, roughly speaking, a measure of the time for which the fluid remembers the flow history. The behavior of non-Newtonian fluids depends crucially on how this time scale relates to other time scales relevant to the flow. If the flow is slow, i.e., on a time scale that is long relative to the memory of the fluid, then memory is unimportant, and the fluid behaves like a Newtonian fluid. On the other hand, memory effects will be crucial if the relaxation time of the fluid exceeds the time scale of the flow. In the extreme case where the time scale of the flow is very short, the fluid will behave like an elastic solid. The ratio of a time scale for the fluid memory to a time scale of the flow is an important dimensionless measure of the importance of elasticity. It is known as the Weissenberg number or the Deborah number. (Deborah was a prophetess in the Old Testament who said that "the mountains flowed before the Lord." Marcus Reiner [67] interpreted this as a statement about the importance of time scales in rheology.)

2.3 Nonlinear models

There are many ways to generalize the linear models discussed above by inclusion of nonlinear terms. The simplest generalization of the Newtonian law is the generalized Newtonian fluid. In this fluid, the extra stress is a function of the velocity gradient, but a nonlinear rather than a linear function:

$$\mathbf{T} = 2\eta(|\mathbf{D}|)\mathbf{D}. \tag{2.7}$$

Here $\mathbf{D}$ is the symmetric part of the velocity gradient,

$$\mathbf{D} = \frac{1}{2}(\nabla\mathbf{v} + (\nabla\mathbf{v})^T), \tag{2.8}$$

and $|\mathbf{D}|$ is the norm

$$|\mathbf{D}| = \sqrt{\sum_{i,j=1}^{3} D_{ij}^2}. \tag{2.9}$$

The viscosity η is a nonlinear function of $|\mathbf{D}|$. The generalized Newtonian fluid cannot account for the effects described in the preceding chapter, but it is often used to model shear flows in situations where the interest is focused on global quantities such as the flow rate in a pipe as a function of pressure drop. Polymeric fluids generally exhibit a decrease in viscosity with increasing shear rate; the decrease can be as much as two or three orders of magnitude. The basic reason for this is the alignment of polymer molecules in the flow direction; to see the plausibility of this leading to a decrease in viscosity, you may compare the "flow behavior" of cooked spaghetti to that of uncooked spaghetti. The prevalence of shear-thinning behavior in real liquids is much to the chagrin of mathematicians who are having a good time proving global existence under the assumption of a viscosity which increases with shear rate.

The nonlinear generalization of Boltzmann's model is a much more complicated affair. We may consider models involving a single integral, but already the question arises whether nonlinearities should be inside the integral, outside the integral, or a combination of both. We may also consider models involving multiple integrals, or there may be nonlinearities which cannot be represented by integrals at all.

A useful concept for formulating nonlinear theories for fluids is the relative deformation gradient. Let $\mathbf{y}(\mathbf{x}, t, s)$ denote the position at time s of the fluid particle which occupies position $\mathbf{x}$ at time t. By definition, we have

$$\frac{\partial}{\partial s}\mathbf{y}(\mathbf{x}, t, s) = \mathbf{v}(\mathbf{y}(\mathbf{x}, t, s), s), \quad \mathbf{y}(\mathbf{x}, t, t) = \mathbf{x}. \tag{2.10}$$

The relative deformation gradient is the matrix defined by

$$F_{ij}(\mathbf{x}, t, s) = \frac{\partial y_i(\mathbf{x}, t, s)}{\partial x_j}. \tag{2.11}$$

In general, the stress in a viscoelastic fluid depends on the history of the relative deformation gradient, i.e., $\mathbf{T}$ is a function of all values of $\mathbf{F}(\mathbf{x}, t, s)$ for times s prior to t:

$$\mathbf{T}(\mathbf{x}, t) = \mathcal{F}(\mathbf{F}(\mathbf{x}, t, s))_{s=-\infty}^{t}. \tag{2.12}$$

The principle of material frame indifference states that stresses arise only in response to deformations of the material; they are not affected if the medium is simply rotated. More precisely, if the material is first rotated and then deformed, the stress ought to be the same, and if the material is first deformed and then rotated, then the stress tensor rotates with the material. That is, if $\mathbf{Q}(s)$ is any orthogonal matrix depending on s, then

$$\begin{aligned} \mathcal{F}(\mathbf{Q}(s)\mathbf{F}(\mathbf{x}, t, s))_{s=-\infty}^{t} &= \mathcal{F}(\mathbf{F}(\mathbf{x}, t, s))_{s=-\infty}^{t}, \\ \mathcal{F}(\mathbf{F}(\mathbf{x}, t, s)\mathbf{Q}^{-1}(t))_{s=-\infty}^{t} &= \mathbf{Q}(t)\mathcal{F}(\mathbf{F}(\mathbf{x}, t, s))_{s=-\infty}^{t}\mathbf{Q}^{-1}(t). \end{aligned} \tag{2.13}$$

(Note that, if the material is rotated at time s, then $\mathbf{F}$ changes to $\mathbf{Q}(s)\mathbf{F}$, and if the material is rotated at time t, then $\mathbf{F}$ changes to $\mathbf{F}\mathbf{Q}^{-1}(t)$.) We shall not enter into a general discussion of restrictions on constitutive theories imposed by material frame indifference or by appropriate formulations of the second law of thermodynamics. These issues are discussed in depth, for instance, in [103]. All the particular models discussed below satisfy these requirements.

A consequence of material frame indifference is that the stress tensor depends on the relative deformation gradient only through the relative Cauchy strain defined by

$$\mathbf{C}(\mathbf{x}, t, s) = \mathbf{F}^T(\mathbf{x}, t, s)\mathbf{F}(\mathbf{x}, t, s). \tag{2.14}$$

That is, we have

$$\mathbf{T}(\mathbf{x}, t) = \mathcal{G}(\mathbf{C}(\mathbf{x}, t, s))_{s=-\infty}^{t}. \tag{2.15}$$

Moreover, this dependence must be isotropic:

$$\mathcal{G}(\mathbf{Q}(t)\mathbf{C}(\mathbf{x}, t, s)\mathbf{Q}^{-1}(t))_{s=-\infty}^{t} = \mathbf{Q}(t)\mathcal{G}(\mathbf{C}(\mathbf{x}, t, s))_{s=-\infty}^{t}\mathbf{Q}^{-1}(t). \tag{2.16}$$

The K-BKZ model [4, 42] is one of the most widely used nonlinear generalizations of Boltzmann's linear model. It can be formulated in terms of a "stored energy" function $W(I_1, I_2, t-s)$, where I_1 and I_2 are the principal invariants of the relative Cauchy strain:

$$I_1 = \operatorname{tr} \mathbf{C}^{-1}(\mathbf{x}, t, s), \quad I_2 = \operatorname{tr} \mathbf{C}(\mathbf{x}, t, s). \tag{2.17}$$

The model has the form

$$\begin{aligned} \mathbf{T}(\mathbf{x}, t) = & \int_{-\infty}^{t} \frac{\partial W(I_1, I_2, t-s)}{\partial I_1}(\mathbf{C}^{-1}(\mathbf{x}, t, s) - \mathbf{I}) \\ & - \frac{\partial W(I_1, I_2, t-s)}{\partial I_2}(\mathbf{C}(\mathbf{x}, t, s) - \mathbf{I})\, ds. \end{aligned} \tag{2.18}$$

Here $\mathbf{I}$ is the identity matrix. The model is motivated by an analogy with nonlinear elasticity. For an isotropic elastic solid, the equilibrium position replaces the position at a prior time. That is, we can define $\mathbf{y}(\mathbf{x},t)$ to be the equilibrium position of the particle which occupies position $\mathbf{x}$ at time t, and we can, as above, define a deformation gradient $\partial\mathbf{y}/\partial\mathbf{x}$ and an associated Cauchy strain and invariants I_1 and I_2. The constitutive laws for an isotropic elastic solid then have the form

$$\mathbf{T}(\mathbf{x},t) = \frac{\partial W(I_1,I_2)}{\partial I_1}(\mathbf{C}^{-1}(\mathbf{x},t) - \mathbf{I}) - \frac{\partial W(I_1,I_2)}{\partial I_2}(\mathbf{C}(\mathbf{x},t) - \mathbf{I}). \qquad (2.19)$$

The K-BKZ model is based on the idea that every prior configuration of the material can be viewed as a "temporary" equilibrium configuration, and the stress is found by superposition of the elastic stresses resulting from all the deformations relative to these temporary equilibrium states. Of course, there is an assumption of linearity here: Although the elastic response associated with a given value of $t-s$ is nonlinear, contributions corresponding to different times s are assumed to superimpose linearly. There are ways of testing such a hypothesis experimentally, and, not surprisingly, the hypothesis is found to be violated in real fluids. This has led to refinements of the K-BKZ model, but, due to the multitude of ways in which integrals and nonlinear functions can be combined, there is no "natural" choice of such refined models. Despite its shortcomings, the K-BKZ model is widely used, and it is a reasonable first step in investigations of integral models.

Differential models are nonlinear extensions of Maxwell's idea of formulating a system of ordinary differential equations which determines the stress in terms of the velocity gradient. The starting point for such models is the upper convected Maxwell (UCM) model, which is a nonlinear modification of Maxwell's linear law which takes account of frame indifference (linear models, apart from the Newtonian fluid, violate frame indifference). The model can also be motivated by molecular theories (see section 2.4). The constitutive equation for the UCM fluid is

$$\frac{\partial\mathbf{T}}{\partial t} + (\mathbf{v}\cdot\nabla)\mathbf{T} - (\nabla\mathbf{v})\mathbf{T} - \mathbf{T}(\nabla\mathbf{v})^T + \lambda\mathbf{T} = 2\mu\mathbf{D}. \qquad (2.20)$$

(Our convention for the gradient of a vector is that the i,j component of $\nabla\mathbf{v}$ is $\partial v_i/\partial x_j$.) The Oldroyd B model has a stress that is the linear superposition of a UCM and a Newtonian contribution. Other popular differential models differ from the UCM by adding additional nonlinearities. A systematic approach for this that includes all possible quadratic terms was given by Oldroyd [58]. Popular models of this type, which are partly motivated by molecular ideas, include the Giesekus model [21], which adds a term proportional to $\mathbf{T}^2$ to the left side of (2.20), the Phan-Thien–Tanner (PTT) model [64], which adds a term proportional to $\mathbf{T}\,\mathrm{tr}\,\mathbf{T}$, and the Johnson–Segalman model [38], which adds a term proportional to $\mathbf{TD}+\mathbf{DT}$. A major reason for adding such terms is that the UCM model overpredicts stresses at large deformation rates (see the next chapter); the other models aim to correct this flaw.

Differential models have been more thoroughly investigated than integral models, since they are easier to implement for numerical simulation. Some models overlap both categories; for instance, the UCM model can be written in the alternative form

$$\mathbf{T}(\mathbf{x},t) = \int_{-\infty}^{t} \mu\lambda e^{-\lambda(t-s)}(\mathbf{C}^{-1}(\mathbf{x},t,s) - \mathbf{I})\,ds, \tag{2.21}$$

i.e., it is a special case of the K-BKZ model. In general, however, systems of nonlinear ODEs cannot be integrated in closed form.

2.4 Molecular theories

"The time has come," the penguin said,
"To speak of many things:
Of flowing macromolecules,
And little beads and springs
That join together into 'chains'
Or even 'stars' or 'rings'."
Robert Byron Bird [5]
(with the assistance of Lewis Carroll)

There is an extensive literature that aims to derive constitutive models from assumptions on the behavior of polymer molecules and their interaction with the flow. There are three basic approaches toward such theories:

1. **Dilute solution theories** treat polymer molecules individually. Each molecule is modeled as a chain of beads and springs or beads and rods. The interaction with the flow results from a hydrodynamic drag that the surrounding fluid exerts on the beads.

2. **Network theories** treat the polymer as a network of springs that are linked at junction points. This type of theory was originally used for solid rubber. In contrast to a solid, where the junctions are permanent, the junctions in a liquid are temporary and form and decay following certain statistical laws. The interaction between the polymer molecule and the flow results from the motion of the junctions, which in the simplest models is assumed to follow the macroscopic motion of the fluid.

3. **Reptation theories** are, in a sense, inbetween these two extremes. The polymer molecules are treated individually. However, in contrast to the dilute solution theories, the motion of the polymer molecule is constrained laterally by a "tube," which represents the other molecules.

We refer to the second volume of [5] for an extensive discussion of molecular theories and references to the literature. In these notes, we shall limit ourselves to a brief discussion of the simplest type of dilute solution theory, the dumbbell model.

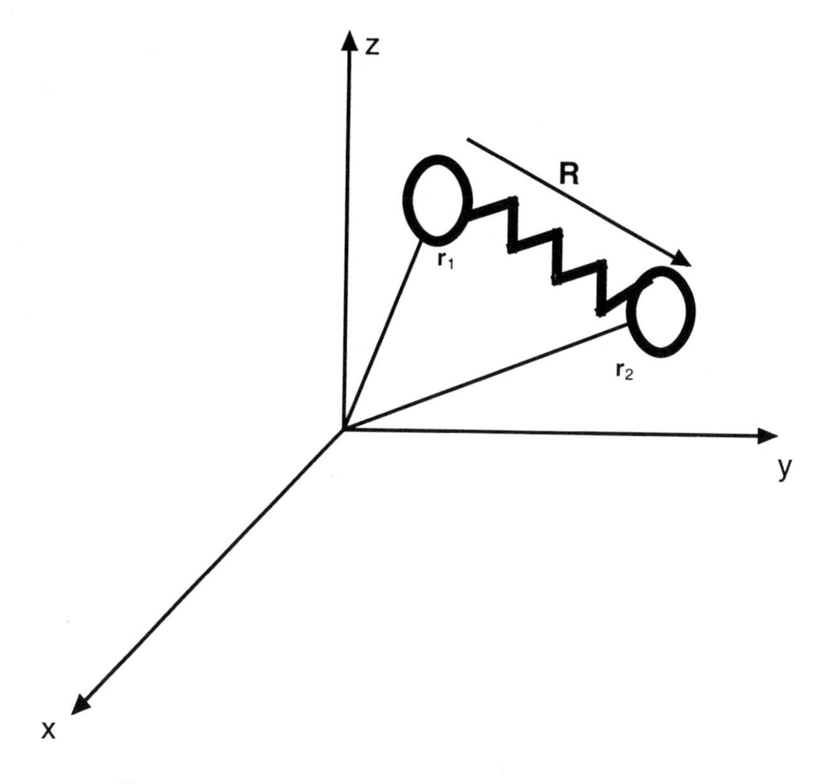

Figure 2.1: Dumbbell model of a polymer molecule.

In a dumbbell model, the polymer molecule is thought of as two beads connected by a spring, as shown in Figure 2.1.

The spring connecting the dumbbells exerts a spring force $\mathbf{F}(\mathbf{R})$. On each of the beads, we have a balance between this spring force, a friction force exerted by the surrounding fluid, and a stochastic force due to Brownian motion (inertial forces at the molecular level can be safely neglected). The friction force, in the simplest model, is assumed to be given by a Stokes drag: $6\pi\eta_s a(\dot{\mathbf{r}}_i - \mathbf{v})$. Here η_s is the solvent viscosity, a is the radius of the bead (we shall assume equal beads), $\dot{\mathbf{r}}_i$ is the velocity of the bead, and $\mathbf{v}$ is the velocity of the surrounding fluid. We shall write ζ for $6\pi\eta_s a$. In practice, η_s is usually not the viscosity of the solvent, but a much larger value representing the viscosity of the polymer solution. Strictly speaking, dilute solution theories only apply if the polymer solution is extremely dilute so that polymer molecules do not see each other and interact only with the solvent. However, if the application of the theories was limited to this situation, they would be useless in practice. As is so often the case in the study of complex phenomena, what is really being done here is to study a limiting case accessible to analysis, but one then hopes that, at least qualitatively, the salient features found in this limiting case will survive beyond the range where the assumptions really apply.

The force balance law for each bead reads

$$\begin{aligned} -\zeta(\dot{\mathbf{r}}_1 - \mathbf{v}(\mathbf{r}_1, t)) + \mathbf{F}(\mathbf{R}) + \mathbf{S}_1 &= \mathbf{0}, \\ -\zeta(\dot{\mathbf{r}}_2 - \mathbf{v}(\mathbf{r}_2, t)) - \mathbf{F}(\mathbf{R}) + \mathbf{S}_2 &= \mathbf{0}. \end{aligned} \tag{2.22}$$

Here $\mathbf{S}_1$ and $\mathbf{S}_2$ denote the stochastic forces due to Brownian motion. We subtract the two equations from each other. Moreover, we make the linear approximation

$$\mathbf{v}(\mathbf{r}_2, t) = \mathbf{v}(\mathbf{r}_1, t) + (\nabla\mathbf{v})(\mathbf{r}_1, t)(\mathbf{r}_2 - \mathbf{r}_1). \tag{2.23}$$

The result is the equation

$$\dot{\mathbf{R}} = \nabla\mathbf{v} \cdot \mathbf{R} - \frac{2}{\zeta}\mathbf{F}(\mathbf{R}) + \frac{1}{\zeta}(\mathbf{S}_2 - \mathbf{S}_1). \tag{2.24}$$

Under reasonable assumptions on the stochastic forces, methods of stochastic differential equations can be used to convert the stochastic differential equation (2.24) to a Fokker–Planck equation. We assume that each macroscopic volume element in space contains a large number of polymer molecules and that the distribution of their connector vectors $\mathbf{R}$ can be described by a probability density $\psi(\mathbf{R}, \mathbf{x}, t)$. For each $\mathbf{x}$ and t, the total probability is equal to 1:

$$\int_{\mathbb{R}^3} \psi(\mathbf{R}, \mathbf{x}, t)\, d\mathbf{R} = 1. \tag{2.25}$$

If the stochastic forces are described by a Wiener process and their magnitude is proportional to kT, then the Fokker–Planck equation is

$$\frac{\partial\psi}{\partial t} + (\mathbf{v} \cdot \nabla_{\mathbf{x}})\psi = \frac{2kT}{\zeta}\Delta_{\mathbf{R}}\psi + \mathrm{div}_{\mathbf{R}}\left[-\nabla\mathbf{v}(\mathbf{x}, t) \cdot \mathbf{R}\psi + \frac{2}{\zeta}\mathbf{F}(\mathbf{R})\psi\right]. \tag{2.26}$$

Here $\Delta_{\mathbf{R}}$ and $\mathrm{div}_{\mathbf{R}}$ indicate differential operators with respect to the variable $\mathbf{R}$.

The solution of the Fokker–Planck equation is a formidable task. Note that ψ is a function of seven variables: three for $\mathbf{R}$, three for $\mathbf{x}$, and one for t. If the dumbbell model is refined by considering chains of beads and springs, the number of variables increases accordingly, and the solution of the Fokker–Planck equation becomes completely impractical. The traditional way to get around this problem is by means of "closure approximations," which make it unnecessary to find the function ψ; instead one has to solve a system of equations involving only a few moments of ψ. The derivation of such a system, however, involves approximations of dubious merit; we shall return to this point later. Recently, numerical simulations that directly attack dumbbell models have been developed. The approach is based on solving not the Fokker–Planck equation, but the original stochastic differential equation with a random force term. This is currently an active and exciting area in numerical simulation of polymeric flows; see [50] for one of the seminal papers.

The connector forces in the polymer molecules contribute to the stress tensor. Each connector carries a force $\mathbf{F}(\mathbf{R})$, and it can be shown that the number of connectors with orientation $\mathbf{R}$ that intersect a plane with normal vector $\mathbf{n}$ is proportional to $n(\mathbf{R}\cdot\mathbf{n})\psi(\mathbf{R},\mathbf{x},t)$, where n is the number density of polymer molecules. As a result, one finds a contribution to the stress tensor that is equal to

$$\mathbf{T}_p(\mathbf{x},t) = n\int_{\mathbb{R}^3} \mathbf{R}\mathbf{F}(\mathbf{R})\psi(\mathbf{R},\mathbf{x},t)\,d\mathbf{R}. \tag{2.27}$$

(The product $\mathbf{RF}$ that appears here is the dyadic product.) The total stress in the polymer solution is then modeled as the sum of this "polymer contribution" and the viscous "solvent stress" $2\eta\mathbf{D}$; again, the viscosity η used in practice is usually not that of the pure solvent.

A variety of force laws for the connector spring are used in the literature. The simplest assumption is a linear force law: $\mathbf{F}(R) = H\mathbf{R}$. If we assume this, then there is good news and good news. The good news is that the Fokker–Planck equation (2.26) can be solved in closed form and the other good news is that we do not have to solve it at all. We shall use the notation $\langle f\rangle$ for

$$\int_{\mathbb{R}^3} f(\mathbf{R})\psi(\mathbf{R},\mathbf{x},t)\,d\mathbf{R}, \tag{2.28}$$

so the stress tensor for a Hookean dumbbell model takes the form

$$\mathbf{T}_p = nH\langle\mathbf{R}\mathbf{R}\rangle. \tag{2.29}$$

Now we multiply the Fokker–Planck equation by $\mathbf{RR}$ and integrate. The result is the equation

$$\frac{\partial\mathbf{C}}{\partial t} + (\mathbf{v}\cdot\nabla)\mathbf{C} = \frac{4kT}{\zeta}\mathbf{I} - \frac{4H}{\zeta}\mathbf{C} + (\nabla\mathbf{v})\mathbf{C} + \mathbf{C}(\nabla\mathbf{v})^T. \tag{2.30}$$

The tensor $\mathbf{C} = \langle\mathbf{RR}\rangle$ is known as the conformation tensor. If we multiply by nH, we get the following equation for the polymer stress $\mathbf{T}_p$:

$$\frac{\partial\mathbf{T}_p}{\partial t} + (\mathbf{v}\cdot\nabla)\mathbf{T}_p - (\nabla\mathbf{v})\mathbf{T}_p - \mathbf{T}_p(\nabla\mathbf{v})^T = \frac{4kTnH}{\zeta}\mathbf{I} - \frac{4H}{\zeta}\mathbf{T}_p. \tag{2.31}$$

With the substitution $\mathbf{T}_p = nkT\mathbf{I} + \tilde{\mathbf{T}}_p$, we find

$$\frac{\partial\tilde{\mathbf{T}}_p}{\partial t} + (\mathbf{v}\cdot\nabla)\tilde{\mathbf{T}}_p - (\nabla\mathbf{v})\tilde{\mathbf{T}}_p - \tilde{\mathbf{T}}_p(\nabla\mathbf{v})^T + \frac{4H}{\zeta}\tilde{\mathbf{T}}_p = 2nkT\mathbf{D}. \tag{2.32}$$

This is precisely the UCM model.

For nonlinear springs, on the other hand, it is not possible to obtain a closed system of equations for the conformation tensor, except by cheating. Of course, we only call it cheating when graduate students do it. When established scientists do it, it is an "approximation" named after the scientist who did the cheating. In the current situation, it is known as the "Peterlin approximation." The force law for a nonlinear spring can be put into the form

$$\mathbf{F}(\mathbf{R}) = \gamma(|\mathbf{R}|^2)\mathbf{R}. \tag{2.33}$$

The Peterlin approximation replaces this by

$$\mathbf{F}(\mathbf{R}) = \gamma(\langle|\mathbf{R}|^2\rangle)\mathbf{R}. \tag{2.34}$$

That is, the length of the spring in the spring constant is replaced by the length of the average spring $\langle|\mathbf{R}|^2\rangle = \mathrm{tr}(\mathbf{C})$. If we do this, then we can derive the following equation for $\mathbf{C}$:

$$\frac{\partial \mathbf{C}}{\partial t} + (\mathbf{v}\cdot\nabla)\mathbf{C} = \frac{4kT}{\zeta}\mathbf{I} - \frac{4\gamma(\mathrm{tr}(\mathbf{C}))}{\zeta}\mathbf{C} + (\nabla\mathbf{v})\mathbf{C} + \mathbf{C}(\nabla\mathbf{v})^T, \tag{2.35}$$

and the stress tensor is given by

$$\mathbf{T}_p = n\gamma(\mathrm{tr}(\mathbf{C}))\mathbf{C}. \tag{2.36}$$

Most rheological studies that claim to be based on molecular dumbbell models are actually based on the Peterlin approximation.

Chapter 3

Behavior in Simple Flows

The first step in evaluating constitutive models is to consider their predictions in a number of simple flows in which the velocity field is known explicitly and it is easy to find the stresses predicted by a given constitutive model. In this chapter, we discuss a few such flows and how some of the simple models behave in these flows.

3.1 Steady simple shear flow

In steady simple shear flow, the flow is two-dimensional and the velocity is unidirectional: $\mathbf{v} = (V(y), 0, 0)$. Consequently, the velocity gradient is the matrix

$$\begin{pmatrix} 0 & V'(y) & 0 \\ 0 & 0 & 0 \\ 0 & 0 & 0 \end{pmatrix}. \tag{3.1}$$

The quantity $V'(y)$ is known as the shear rate.

If $\mathbf{x} = (x, y, z)$ denotes the position of a fluid particle at time t, then its position at time s is $(x - V(y)(t-s), y, z)$, so that the relative deformation gradient is given by

$$\mathbf{F}(\mathbf{x}, t, s) = \begin{pmatrix} 1 & -V'(y)(t-s) & 0 \\ 0 & 1 & 0 \\ 0 & 0 & 1 \end{pmatrix} \tag{3.2}$$

and the relative Cauchy strain is

$$\mathbf{C}(\mathbf{x}, t, s) = \begin{pmatrix} 1 & -V'(y)(t-s) & 0 \\ -V'(y)(t-s) & 1 + V'(y)^2(t-s)^2 & 0 \\ 0 & 0 & 1 \end{pmatrix}. \tag{3.3}$$

Thus $\mathbf{C}(\mathbf{x}, t, s)$ is completely determined by the shear rate $\kappa = V'(y)$, and hence the stress tensor is a function of κ; see (2.15). Moreover, we have the invariance given by (2.16), and if we take $\mathbf{Q}(t)$ to be a 180° rotation about the z-axis, we find that $\mathbf{C}(\mathbf{x},t,s)$ as given by (3.3) is preserved:

$$\mathbf{Q}(t)\mathbf{C}(\mathbf{x},t,s)\mathbf{Q}^{-1}(t) = \mathbf{C}(\mathbf{x},t,s). \tag{3.4}$$

Hence the stress $\mathbf{T}$ must also be preserved under this transformation, and consequently the components T_{13} and T_{23} are zero. The stress tensor in simple shear flow has the form

$$\mathbf{T} = \begin{pmatrix} T_{11}(\kappa) & T_{12}(\kappa) & 0 \\ T_{12}(\kappa) & T_{22}(\kappa) & 0 \\ 0 & 0 & T_{33}(\kappa) \end{pmatrix}. \tag{3.5}$$

We also note that because of the presence of an undetermined pressure in an incompressible fluid, the diagonal components of the extra stress tensor have physical meaning only modulo an arbitrary constant; hence it is natural to consider their differences. This leads to the three viscometric functions

$$T_{12}(\kappa) = \eta(\kappa)\kappa, \quad T_{11}(\kappa) - T_{22}(\kappa) = N_1(\kappa), \quad T_{22}(\kappa) - T_{33}(\kappa) = N_2(\kappa). \tag{3.6}$$

Here $\eta(\kappa)$ is called the viscosity, and N_1 and N_2 are called the first and second normal stress differences. In a Newtonian fluid, $\eta(\kappa)$ is constant and N_1 and N_2 are zero.

We shall now compare the viscometric behavior of some simple differential models, starting with the UCM model. If we restrict (2.20) to steady simple shear flow, we find

$$\lambda T_{11} - 2\kappa T_{12} = 0, \quad \lambda T_{12} - \kappa T_{22} = \mu\kappa, \quad T_{22} = T_{33} = 0. \tag{3.7}$$

Consequently, $\eta(\kappa) = \mu/\lambda$, $N_1(\kappa) = 2\mu\kappa^2/\lambda^2$, $N_2(\kappa) = 0$. Hence the viscosity is constant, the first normal stress difference is a quadratic function, and the second normal stress difference is zero. In real polymeric fluids, the viscosity typically decreases with increasing shear rate, the first normal stress difference grows quadratically at low shear rates but then grows more slowly as the shear rate increases further, and the second normal stress difference is negative, but much smaller in magnitude than the first (a typical value of N_2/N_1 is -0.1). More refined models than the UCM aim at capturing the shear thinning and, in some cases, the presence of a nonzero second normal stress difference.

The PTT model [64] has the constitutive equation

$$\frac{\partial \mathbf{T}}{\partial t} + (\mathbf{v}\cdot\nabla)\mathbf{T} - (\nabla\mathbf{v})\mathbf{T} - \mathbf{T}(\nabla\mathbf{v})^T + \lambda\mathbf{T} + \nu(\operatorname{tr}\mathbf{T})\mathbf{T} = 2\mu\mathbf{D}. \tag{3.8}$$

Here ν is assumed positive. For steady simple shear flow, this specializes to

$$\begin{aligned} -2\kappa T_{12} + \lambda T_{11} + \nu(T_{11}+T_{22}+T_{33})T_{11} &= 0, \\ -\kappa T_{22} + \lambda T_{12} + \nu(T_{11}+T_{22}+T_{33})T_{12} &= \mu\kappa, \\ \lambda T_{22} + \nu(T_{11}+T_{22}+T_{33})T_{22} &= 0, \\ \lambda T_{33} + \nu(T_{11}+T_{22}+T_{33})T_{33} &= 0. \end{aligned} \tag{3.9}$$

From the last two equations, we find $T_{22} = T_{33} = 0$, and we are left with

$$\begin{aligned} -2\kappa T_{12} + \lambda T_{11} + \nu T_{11}^2 &= 0, \\ \lambda T_{12} + \nu T_{11} T_{12} &= \mu\kappa. \end{aligned} \tag{3.10}$$

We can eliminate T_{11} from the second equation and substitute into the first. The result is the cubic equation

$$2\nu T_{12}^3 + \lambda\mu T_{12} = \mu^2\kappa. \tag{3.11}$$

This shows that the shear stress T_{12} is proportional to κ at low shear rates, but proportional to $\kappa^{1/3}$ at high shear rates. The first normal stress difference $N_1 = T_{11}$ satisfies

$$\lambda T_{11} + \nu T_{11}^2 = 2\kappa T_{12}, \tag{3.12}$$

and hence T_{11} is proportional to κ^2 at low shear rates, but proportional to $\kappa^{2/3}$ at high shear rates.

For the Johnson–Segalman model [38], we have the constitutive relation

$$\frac{\partial \mathbf{T}}{\partial t} + (\mathbf{v} \cdot \nabla)\mathbf{T} - (\nabla \mathbf{v})\mathbf{T} - \mathbf{T}(\nabla \mathbf{v})^T + \lambda \mathbf{T} + \nu(\mathbf{TD} + \mathbf{DT}) = 2\mu \mathbf{D}. \tag{3.13}$$

In simple shear flow, this specializes to

$$\begin{aligned} -2\kappa T_{12} + \lambda T_{11} + \nu\kappa T_{12} &= 0, \\ -\kappa T_{22} + \lambda T_{12} + \nu\kappa \frac{T_{11} + T_{22}}{2} &= \mu\kappa, \\ \lambda T_{22} + \nu\kappa T_{12} &= 0, \\ \lambda T_{33} &= 0. \end{aligned} \tag{3.14}$$

We find $T_{33} = 0$, $T_{22} = -\nu\kappa T_{12}/\lambda$, and $T_{11} = (2 - \nu)\kappa T_{12}/\lambda$. The ratio of the normal stress differences is therefore

$$\frac{N_2(\kappa)}{N_1(\kappa)} = -\frac{\nu}{2}. \tag{3.15}$$

If the sign of the normal stresses is to be consistent with experiments ($N_1 > 0$, $N_2 < 0$), then only values of ν between 0 and 2 need to be considered. Moreover, the analysis of rod climbing [41] has shown that, at least in the limit of low shear rates, $|N_2/N_1|$ should be less than 1/4; this restricts ν to be less than 1/2. After eliminating T_{11} and T_{22} from the equations, we find a single equation for T_{12}, which yields the solution

$$T_{12} = \frac{\mu\lambda\kappa}{\nu(2 - \nu)\kappa^2 + \lambda^2}. \tag{3.16}$$

We see that at low shear rates the shear stress is proportional to κ, but then the shear stress reaches a maximum at $\kappa = \lambda/\sqrt{\nu(2 - \nu)}$ and then decreases proportional to $1/\kappa$ for large shear rates. The Johnson–Segalman model is therefore very strongly shear thinning. If the stress is a linear combination of the

Johnson–Segalman model and a Newtonian contribution, it is possible to have a shear stress that first increases with shear rate, then decreases, and ultimately increases again. Hence it is possible to have two different shear rates at the same shear stress. This has been suggested as an explanation for formation of shear bands and spurt.

The constitutive relation for the Giesekus model is

$$\frac{\partial \mathbf{T}}{\partial t} + (\mathbf{v}\cdot\nabla)\mathbf{T} - (\nabla\mathbf{v})\mathbf{T} - \mathbf{T}(\nabla\mathbf{v})^T + \lambda\mathbf{T} + \nu\mathbf{T}^2 = 2\mu\mathbf{D}, \tag{3.17}$$

and the special case of steady simple shear flow reduces to

$$\begin{aligned} -2\kappa T_{12} + \lambda T_{11} + \nu(T_{11}^2 + T_{12}^2) &= 0, \\ -\kappa T_{22} + \lambda T_{12} + \nu(T_{11} + T_{22})T_{12} &= \mu\kappa, \\ \lambda T_{22} + \nu(T_{12}^2 + T_{22}^2) &= 0, \\ \lambda T_{33} + \nu T_{33}^2 &= 0. \end{aligned} \tag{3.18}$$

The physically relevant solution of the last equation is $T_{33} = 0$. The remaining three equations, however, are somewhat more complicated than in the preceding examples. We can eliminate T_{11} and T_{22} and obtain an equation relating the shear stress to the shear rate. The outcome of this calculation is

$$\begin{aligned} &\kappa^2(\lambda\mu - \mu^2\nu - \nu T_{12}^2)^2 \\ &+\kappa T_{12}(-\lambda^3\mu + \lambda^2\mu^2\nu - \lambda^2\nu T_{12}^2 + 8\lambda\mu\nu^2 T_{12}^2 - 8\mu^2\nu^3 T_{12}^2) \\ &+\nu T_{12}^2(\lambda^3\mu - \lambda^2\mu^2\nu + \lambda^2\nu T_{12}^2 - 4\lambda\mu\nu^2 T_{12}^2 + 4\mu^2\nu^3 T_{12}^2) = 0. \end{aligned} \tag{3.19}$$

We can look at this as a quadratic equation for the shear rate κ for given shear stress T_{12}. The discriminant of the quadratic equation is

$$D = (\lambda - 2\mu\nu)^2 T_{12}^2(\lambda^2 - 4\nu^2 T_{12}^2)(\lambda\mu - \mu^2\nu + \nu T_{12}^2)^2. \tag{3.20}$$

We find that there are no solutions for κ if $|T_{12}| > \lambda/(2\nu)$, i.e., there is a limiting value for the shear stress. If $|T_{12}| < \lambda/(2\nu)$, then there are usually two solutions. However, these two solutions coincide if $\nu = \lambda/(2\mu)$, $T_{12} = 0$, or $\lambda\mu - \mu^2\nu + \nu T_{12}^2 = 0$. Next we look at limiting cases. In the limit $\kappa \to \infty$, the shear stress approaches the value which makes the quadratic term in (3.19) equal to zero:

$$\lambda\mu - \mu^2\nu - \nu T_{12}^2 = 0. \tag{3.21}$$

Evidently, this equation has solutions only if $\nu \leq \lambda/\mu$. If ν is larger than this value, then there are no solutions for high shear rates, i.e., the model has a limiting shear rate, beyond which steady shear flow cannot exist. We shall from now on assume that $\nu < \lambda/\mu$. If κ and T_{12} are small, then the balance of dominant terms in (3.19) yields

$$\mu^2(\lambda - \mu\nu)^2\kappa^2 - T_{12}\kappa\lambda^2\mu(\lambda - \mu\nu) + \nu T_{12}^2\lambda^2\mu(\lambda - \mu\nu) = 0, \tag{3.22}$$

which can be factored as

$$\mu(\lambda - \mu\nu)(\lambda T_{12} - \mu\kappa)(\nu\lambda T_{12} + (\nu\mu - \lambda)\kappa) = 0. \tag{3.23}$$

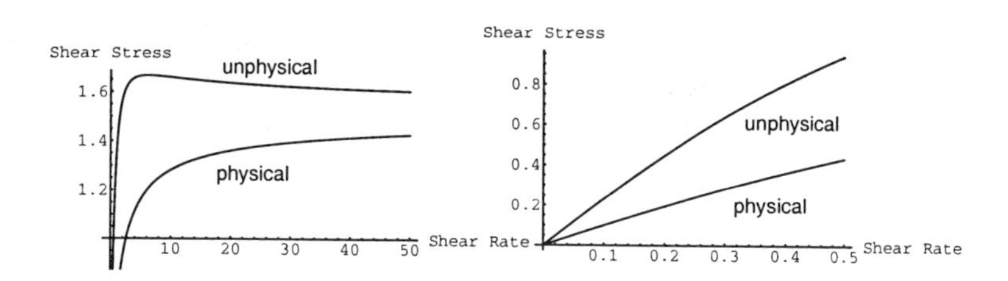

Figure 3.1: Shear stress vs. shear rate for the Giesekus model if $\nu < \lambda/(2\mu)$.

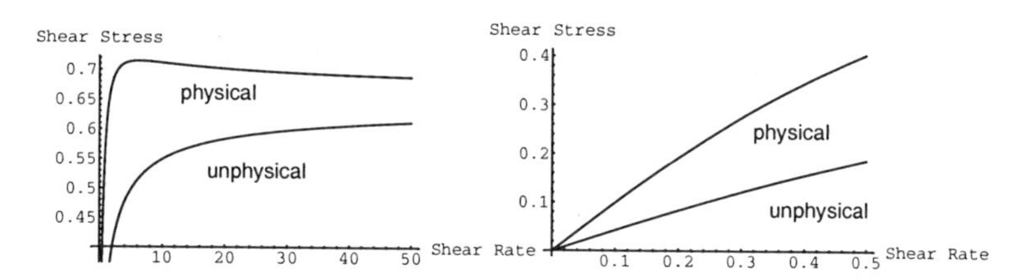

Figure 3.2: Shear stress vs. shear rate for the Giesekus model if $\nu > \lambda/(2\mu)$.

We find the two possible solutions $T_{12} = \mu\kappa/\lambda$ and $T_{12} = (\lambda - \nu\mu)\kappa/(\nu\lambda)$. The second of these solutions can be shown to lead to a nonzero normal stress difference in the limit of zero shear rate and is ruled out as unphysical.

The overall behavior of the shear stress vs. shear rate relation is as follows: if $\nu < \lambda/(2\mu)$, then the unphysical solution has a larger shear stress than the physical solution. At some value of the shear rate, the unphysical solution reaches the shear stress maximum $\lambda/(2\nu)$. As the shear rate increases further, the shear stress on the unphysical branch decreases, while the shear stress on the physical branch continues to increase. As $\kappa \to \infty$, the shear stress on both branches approaches the common limit $\sqrt{(\lambda - \mu\nu)\mu/\nu}$. An example is shown in Figure 3.1. The first part of the figure shows the behavior at high shear rates, the second shows the behavior at low shear rates.

If $\nu = \lambda/(2\mu)$, the two branches coincide, and if $\nu > \lambda/(2\mu)$, the larger value of shear stress and the stress maximum occur on the physical branch, as shown in Figure 3.2.

We see that the Giesekus model is more strongly shear thinning than the PTT model, but not as much so as the Johnson–Segalman model. If $\nu > \lambda/(2\mu)$, then, as in the Johnson–Segalman model, the shear stress is a nonmonotone function of shear rate. Like the Johnson–Segalman model, the Giesekus model has a nonzero second normal stress difference.

3.2 Steady viscometric flows

We can generate a steady simple shear flow as described in the preceding section by putting the fluid between two infinite parallel plates and sliding one of the plates with constant speed. In practice, however, infinite plates are hard to come by. Fortunately, however, there is a whole class of flows that are equivalent to simple shear flows, and some of these flows are easier to produce experimentally. We shall call a flow a steady viscometric flow if the relative deformation gradient has the form

$$\mathbf{F}(\mathbf{x},t,s) = \mathbf{Q}(t-s,\mathbf{x})(1-\kappa(\mathbf{x})(t-s)\mathbf{N}(\mathbf{x})), \tag{3.24}$$

where $\mathbf{Q}$ is an orthogonal matrix, $\kappa(\mathbf{x})$ is a constant, and $\mathbf{N}$ is a matrix with the properties

$$\mathbf{N}^2 = \mathbf{0}, \quad |\mathbf{N}| = \sqrt{\sum_{i,j=1}^{3} N_{ij}^2} = 1. \tag{3.25}$$

If the relative deformation gradient has the form given by (3.24), then the relative Cauchy strain is

$$\mathbf{C}(\mathbf{x},t,s) = \mathbf{I} - \kappa(\mathbf{x})(t-s)(\mathbf{N}(\mathbf{x}) + \mathbf{N}^T(\mathbf{x})) + \kappa(\mathbf{x})^2(t-s)^2\mathbf{N}^T(\mathbf{x})\mathbf{N}(\mathbf{x}). \tag{3.26}$$

It can be shown that every matrix $\mathbf{N}(\mathbf{x})$ that satisfies (3.25) is of the form

$$\mathbf{N}(\mathbf{x}) = \mathbf{Q}(\mathbf{x}) \begin{pmatrix} 0 & 1 & 0 \\ 0 & 0 & 0 \\ 0 & 0 & 0 \end{pmatrix} \mathbf{Q}^{-1}(\mathbf{x}), \tag{3.27}$$

where $\mathbf{Q}(\mathbf{x})$ is an orthogonal matrix. As a consequence, $\mathbf{C}(\mathbf{x},t,s)$ becomes

$$\mathbf{Q}(\mathbf{x}) \begin{pmatrix} 1 & -\kappa(\mathbf{x})(t-s) & 0 \\ -\kappa(\mathbf{x})(t-s) & 1+\kappa(\mathbf{x})^2(t-s)^2 & 0 \\ 0 & 0 & 1 \end{pmatrix} \mathbf{Q}(\mathbf{x})^T. \tag{3.28}$$

Except for a rotation, this is exactly the same as in simple shear flow with shear rate $\kappa(\mathbf{x})$, and consequently the stress is also the same as in simple shear flow except for a rotation.

The most important viscometric flows used in experiments are torsional flows. In such flows, particles move on circles around a given axis, but the rotation speed may vary with position. We use the notation $\mathbf{x} = (x,y,z)$ for a point in space, and we shall use capital letters (X,Y,Z) for the components of $\mathbf{y}(\mathbf{x},t,s)$. If the axis of rotation is the z-axis and ω is the rotation speed, then we have

$$\begin{aligned} X &= \cos(\omega(r,z)(t-s))x + \sin(\omega(r,z)(t-s))y, \\ Y &= -\sin(\omega(r,z)(t-s))x + \cos(\omega(r,z)(t-s))y, \\ Z &= z. \end{aligned} \tag{3.29}$$

Here r denotes $\sqrt{x^2+y^2}$, i.e., the distance from the axis of rotation. The deformation gradient is

$$\begin{aligned}
\mathbf{F}(\mathbf{x},t,s) &= \begin{pmatrix} C & S & 0 \\ -S & C & 0 \\ 0 & 0 & 1 \end{pmatrix} \\
&\quad +(t-s)\omega_r \begin{pmatrix} -\frac{x^2}{r}S+\frac{xy}{r}C & -\frac{xy}{r}S+\frac{y^2}{r}C & 0 \\ -\frac{x^2}{r}C-\frac{xy}{r}S & -\frac{xy}{r}C-\frac{y^2}{r}S & 0 \\ 0 & 0 & 0 \end{pmatrix} \\
&\quad +(t-s)\omega_z \begin{pmatrix} 0 & 0 & -xS+yC \\ 0 & 0 & -xC-yS \\ 0 & 0 & 0 \end{pmatrix}. \qquad (3.30)
\end{aligned}$$

Here we have written C for $\cos(\omega(t-s))$ and S for $\sin(\omega(t-s))$. We can put the equation above in the form $\mathbf{F}(\mathbf{x},t,s) = \mathbf{Q}(t-s,\mathbf{x})(\mathbf{I}+(t-s)\mathbf{M}(\mathbf{x}))$, where

$$\mathbf{Q}(t-s,x) = \begin{pmatrix} C & S & 0 \\ -S & C & 0 \\ 0 & 0 & 1 \end{pmatrix} \qquad (3.31)$$

and

$$\mathbf{M}(\mathbf{x}) = \begin{pmatrix} \omega_r\frac{xy}{r} & \omega_r\frac{y^2}{r} & \omega_z y \\ -\omega_r\frac{x^2}{r} & -\omega_r\frac{xy}{r} & -\omega_z x \\ 0 & 0 & 0 \end{pmatrix}. \qquad (3.32)$$

It is easily verified that $\mathbf{M}(\mathbf{x})^2 = \mathbf{0}$ and

$$\kappa(\mathbf{x}) := |\mathbf{M}(\mathbf{x})| = r\sqrt{\omega_r^2+\omega_z^2}. \qquad (3.33)$$

Consequently, the flow is viscometric with shear rate $\kappa(\mathbf{x})$.

Torsional flows of this type include the flow between concentric rotating cylinders, the flow between parallel rotating plates, and the flow between a cone and a plate.

3.3 Steady elongational flows

Elongational flows are flows in which the fluid undergoes a stretching motion. We can distinguish uniaxial, planar, and biaxial extension. Steady uniaxial extension has a velocity field of the form $\mathbf{v} = (\kappa x, -\kappa y/2, -\kappa z/2)$, planar extension has a velocity of the form $\mathbf{v} = (\kappa x, -\kappa y, 0)$, and biaxial extension is the same as uniaxial extension with a negative κ.

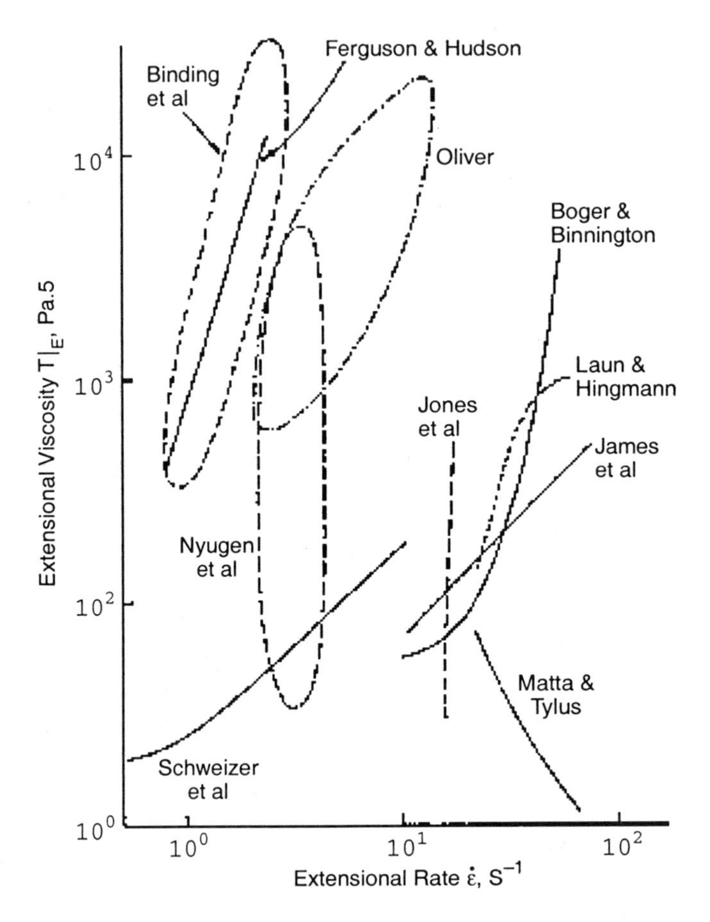

Figure 3.3: Comparison of elongational viscosity measurements for the same fluid (Figure 2.1 of [37]). Reprinted from D.F. James and K. Waters, A critical appraisal of available methods for the measurement of extensional properties of mobile systems, in *Techniques in Rheological Measurement*, A.A. Collyci, ed., pp. 33–53, with kind permission from Kluwer Academic Publishers, ©1993 by Chapman and Hall.

While it is easy to postulate such a velocity field theoretically, it is another matter to create it experimentally. A number of experiments aim at approximating steady elongational flows, and none is truly satisfactory. As a consequence, data on elongational stresses are harder to come by and much more uncertain than shear data. Sometimes the best that can be said about elongational viscosities "measured" by different methods is that all the points lie in a plane. See Figure 3.3 (from [37]) for an example. Numerical simulations can be used to gain a better understanding of the true behavior of flows that were meant to approximate steady elongational flow but are actually much more complex.

In this section, we shall review the elongational behavior of the same models that we discussed in shear flow above. We shall limit our consideration to uniaxial elongation. The stress in such a flow is diagonal, with $T_{22} = T_{33}$, and we are interested in the stress difference $T_{11} - T_{22}$ as a function of κ.

In a Newtonian fluid, we have $T_{11} = 2\eta\kappa$, $T_{22} = -\eta\kappa$, so the difference is $3\eta\kappa$. The quantity $(T_{11} - T_{22})/\kappa$ is called the elongational viscosity, and the ratio of elongational to shear viscosity is called the Trouton ratio. Hence the Trouton ratio for a Newtonian fluid is 3. For the UCM fluid, we have

$$T_{11} = \frac{2\mu\kappa}{\lambda - 2\kappa}, \quad T_{22} = -\frac{\mu\kappa}{\lambda + \kappa}, \tag{3.34}$$

hence

$$T_{11} - T_{22} = \frac{3\mu\kappa\lambda}{(\lambda + \kappa)(\lambda - 2\kappa)}. \tag{3.35}$$

At low elongation rates, the Trouton ratio is 3 as in a Newtonian fluid, but then the elongational viscosity increases rapidly with elongation rate and actually becomes infinite at $\kappa = \lambda/2$. Steady elongational flows with a larger elongation rate are impossible; although a solution of the equations with a negative $T_{11} - T_{22}$ exists, this solution is unattainable; it cannot be reached if the fluid is deformed starting from equilibrium. Polymeric fluids do indeed show an increase in elongational viscosity with elongation rate. In polymer solutions, this increase can well be by three or four orders of magnitude; in polymer melts, it seldom reaches even one order of magnitude. Hence the prediction of a limiting elongation rate where the elongational viscosity becomes infinite is reasonable for some polymeric fluids, but not for all of them.

For the Johnson–Segalman model, we find

$$T_{11} = \frac{2\mu\kappa}{\lambda - 2\kappa + 2\kappa\nu}, \quad T_{22} = -\frac{\mu\kappa}{\lambda + \kappa - \kappa\nu}. \tag{3.36}$$

As long as $\nu < 1$, the behavior is similar to the UCM model. Although the Johnson–Segalman model is the most strongly shear thinning of all the models considered here, it does not change the elongational behavior in a qualitative fashion.

For the Giesekus model, we find

$$\begin{aligned} -2\kappa T_{11} + \lambda T_{11} + \nu T_{11}^2 &= 2\mu\kappa, \\ \kappa T_{22} + \lambda T_{22} + \nu T_{22}^2 &= -\mu\kappa. \end{aligned} \tag{3.37}$$

For physically relevant solutions, T_{11} is positive and $-\mu < T_{22} < 0$. For any positive κ, this determines a unique value of T_{11} and T_{22}. The elongational stress no longer reaches infinity at a finite κ; instead, the growth of T_{11} for $\kappa \to \infty$ is proportional to κ, with the Trouton ratio approaching the limit $2\lambda/(\mu\nu)$.

For the PTT model, we have

$$\begin{aligned} -2\kappa T_{11} + \lambda T_{11} + \nu(T_{11} + 2T_{22})T_{11} &= 2\mu\kappa, \\ \kappa T_{22} + \lambda T_{22} + \nu(T_{11} + 2T_{22})T_{22} &= -\mu\kappa. \end{aligned} \tag{3.38}$$

We can eliminate T_{22} and then solve for κ as a function of T_{11}. The result of this calculation is

$$\kappa = \frac{T_{11}(2\lambda\mu + 3\lambda T_{11} + 3\nu T_{11}^2)}{4\mu^2 + 10\mu T_{11} + 6T_{11}^2}. \tag{3.39}$$

In the limit of low elongation rates, we recover the Newtonian result that $T_{11} = 2\mu\kappa/\lambda$, while at high elongation rates, we have $T_{11} = 2\kappa/\nu$. The limiting value of the elongational viscosity is therefore the same as for the Giesekus model.

Chapter 4

Existence Theory

For the Navier–Stokes equations, there is a well-developed theory of existence, qualitative dynamics, and numerical approximation, even though the famous question of global existence in three space dimensions is still outstanding. I refer to [48], [101], and [102] for an introduction to this field. For viscoelastic fluids, virtually everything that is known about basic existence questions has been established in the past twenty years, and, if we discount models specifically cooked up by mathematicians to make existence proofs easy, the results are still fragmentary. In particular, there is as yet no understanding of global issues, and all known existence results (except for some one-dimensional flow problems) are based in some way or other on small perturbations. There are basically four types of existence results:

1. Results on existence locally in time for initial value problems. The solution in this case is a small perturbation of the initial data.

2. Results on global time existence and asymptotic decay if the initial condition is a small perturbation of the rest state.

3. Results on existence of steady flows which are a small perturbation of the rest state.

4. Results on existence of steady flows which are a small perturbation of a Newtonian flow.

For an introduction to existence results in viscoelastic flows and a review of results in the field up to 1986 I refer to [68]. A more recent review of the area was given in an article by Guillopé and Saut [30].

4.1 Initial value problems

Complete proofs are long and technical, and in this chapter, we shall confine ourselves to a review of the state of the art and a basic outline of the main ideas. We begin with a discussion of local time existence results for initial value problems. We first discuss differential constitutive models; the result we cite is from [76]. Here we consider the system of equations consisting of the momentum equation

$$\rho\Big(\frac{\partial \mathbf{v}}{\partial t} + (\mathbf{v}\cdot\nabla)\mathbf{v}\Big) = \operatorname{div}\mathbf{T} - \nabla p + \mathbf{f}, \tag{4.1}$$

the incompressibility condition

$$\operatorname{div}\mathbf{v} = 0, \tag{4.2}$$

and a constitutive relation of the form

$$\Big(\frac{\partial}{\partial t} + (\mathbf{v}\cdot\nabla)\Big)T_{ij} = \sum_{k,l} A_{ijkl}(\mathbf{T})\frac{\partial v_k}{\partial x_l} + g_{ij}(\mathbf{T}). \tag{4.3}$$

All the differential constitutive models discussed in Chapter 2 are of this form. The equations are to be solved on a smooth bounded domain $\Omega \subset \mathbb{R}^3$, subject to the boundary condition $\mathbf{v} = \mathbf{0}$ on $\partial\Omega$ and initial conditions

$$\mathbf{v}(\mathbf{x},0) = \mathbf{v}_0(\mathbf{x}),\ \ \mathbf{T}(\mathbf{x},0) = \mathbf{T}_0(\mathbf{x}). \tag{4.4}$$

The existence theorem basically states that for sufficiently nice data there is a smooth solution of the initial value problem on some time interval. Specifically, we make the following smoothness assumptions:

1. The domain Ω is bounded and $\partial\Omega$ is of class C^5.
2. The functions A_{ijkl} and g_{ij} are of class C^4.
3. The initial data satisfy $\mathbf{v}_0 \in H^4(\Omega)$ and $\mathbf{T}_0 \in H^4(\Omega)$ (we commit the slight abuse of using the same notation for function spaces regardless of whether the functions are scalar, vector, or tensor valued).
4. The prescribed body force satisfies

$$\mathbf{f} \in \bigcap_{k=0}^{4} W^{k,1}([0,T]; H^{4-k}(\Omega)). \tag{4.5}$$

In addition to smoothness, we need compatibility conditions. We require that $\mathbf{v}_0$ be divergence free and vanish on $\partial\Omega$. For given $\mathbf{v}_0$ and $\mathbf{T}_0$, we can determine an initial value of

$$\rho\frac{\partial \mathbf{v}}{\partial t} + \nabla p = -\rho(\mathbf{v}_0\cdot\nabla)\mathbf{v}_0 + \operatorname{div}\mathbf{T}_0 + \mathbf{f}(\mathbf{x},0) \tag{4.6}$$

from the momentum equation. Since $\partial\mathbf{v}/\partial t$ should be divergence free, we can use the Hodge decomposition to find the initial value of $\partial\mathbf{v}/\partial t$. We can repeat this procedure for higher time derivatives if we consider time derivatives of the momentum equation. Our compatibility condition is that the initial values of

$\partial \mathbf{v}/\partial t$, $\partial^2 \mathbf{v}/\partial t^2$, and $\partial^3 \mathbf{v}/\partial t^3$ (which are divergence free by definition) must also vanish on the boundary $\partial\Omega$.

The most important assumption is a structure condition on the coefficients A_{ijkl}. First of all, it can be shown that frame indifference requires that

$$A_{ijkl}(\mathbf{T}) = \frac{1}{2}(\delta_{ik}T_{lj} - \delta_{il}T_{kj} - T_{ik}\delta_{lj} + T_{il}\delta_{kj}) + B_{ijkl}(\mathbf{T}), \tag{4.7}$$

where B_{ijkl} is symmetric with respect to k and l. Since the stress tensor is symmetric, it is of course also symmetric with respect to i and j. We shall assume the additional symmetry

$$B_{ijkl} = B_{klij}. \tag{4.8}$$

In addition to this symmetry, a strong ellipticity condition is required. This condition is stated in terms of

$$C_{ijkl} = A_{ijkl} - T_{il}\delta_{kj}; \tag{4.9}$$

it is easily verified that C_{ijkl} also satisfies (4.8). Strong ellipticity is the requirement that

$$\sum_{i,j,k,l} C_{ijkl}(\mathbf{T})\zeta_i\zeta_k\eta_j\eta_l \geq \kappa(\mathbf{T})|\zeta|^2|\eta|^2 \ \forall \zeta, \eta \in \mathbb{R}^3, \tag{4.10}$$

where $\kappa(\mathbf{T}) > 0$. Under the assumptions just outlined, we can prove the following theorem.

Theorem 1 *There exists a $T' > 0$ for which the initial boundary value problem has a unique solution with the regularity*

$$\mathbf{v} \in \bigcap_{k=0}^{4} C^k([0,T'], H^{4-k}(\Omega)), \ \mathbf{T} \in \bigcap_{k=0}^{3} C^k([0,T'], H^{3-k}(\Omega)). \tag{4.11}$$

The proof of the theorem is based on an iterative construction of the solution. The iteration alternates between finding a new stress for given velocity and finding a new velocity for given stress. The determination of the new stress is simply based on integrating the differential constitutive law, but the formulation of a problem for the velocity requires a little preparation. Toward this end, we apply the operation

$$\frac{\partial}{\partial t} + (\mathbf{v} \cdot \nabla) + (\nabla \mathbf{v})^T \tag{4.12}$$

to the momentum equation. We note that

$$\left(\frac{\partial}{\partial t} + (\mathbf{v} \cdot \nabla) + (\nabla \mathbf{v})^T\right) \nabla p = \nabla \left(\frac{\partial p}{\partial t} + (\mathbf{v} \cdot \nabla)p\right) \tag{4.13}$$

and

$$\begin{aligned}\left(\frac{\partial}{\partial t} + (\mathbf{v} \cdot \nabla) + (\nabla \mathbf{v})^T\right) \operatorname{div} \mathbf{T} = {} & \operatorname{div} \left(\frac{\partial \mathbf{T}}{\partial t} + (\mathbf{v} \cdot \nabla)\mathbf{T}\right) \\ & + (\nabla \mathbf{v})^T(\operatorname{div} \mathbf{T}) - ((\nabla \mathbf{v})^T \cdot \nabla) \cdot \mathbf{T}.\end{aligned} \tag{4.14}$$

We can use the constitutive equation to replace

$$\frac{\partial \mathbf{T}}{\partial t} + (\mathbf{v} \cdot \nabla)\mathbf{T}. \tag{4.15}$$

The result is an equation of the form

$$\begin{aligned} \rho\Big(\frac{\partial}{\partial t} + (\mathbf{v} \cdot \nabla)\Big)^2 v_i &= -\frac{\partial q}{\partial x_i} + \sum_{j,k,l} C_{ijkl}(\mathbf{T}) \frac{\partial^2 v_k}{\partial x_j \partial x_l} \\ &\quad + h_i\left(\mathbf{v}, \nabla \mathbf{v}, \frac{\partial \mathbf{v}}{\partial t}, \mathbf{T}, \nabla \mathbf{T}, \mathbf{f}, \nabla \mathbf{f}, \frac{\partial \mathbf{f}}{\partial t}\right). \end{aligned} \tag{4.16}$$

Here we have set $q = \partial p/\partial t + (\mathbf{v} \cdot \nabla)p$.

We can now describe the iteration that leads to the construction of a solution. For a given velocity iterate $\mathbf{v}^n$, we first determine a stress field $\mathbf{T}^n$ by solving the constitutive relation

$$\Big(\frac{\partial}{\partial t} + (\mathbf{v}^n \cdot \nabla)\Big) T_{ij}^n = \sum_{k,l} A_{ijkl}(\mathbf{T}^n) \frac{\partial v_k^n}{\partial x_l} + g_{ij}(\mathbf{T}^n) \tag{4.17}$$

subject to the initial condition

$$\mathbf{T}^n(\mathbf{x}, 0) = \mathbf{T}_0(\mathbf{x}). \tag{4.18}$$

Then we determine a new velocity iterate from the equation

$$\begin{aligned} \rho\Big(\frac{\partial}{\partial t} + (\mathbf{v}^n \cdot \nabla)\Big)^2 v_i^{n+1} &= -\frac{\partial q^{n+1}}{\partial x_i} + \sum_{j,k,l} C_{ijkl}(\mathbf{T}^n) \frac{\partial^2 v_k^{n+1}}{\partial x_j \partial x_l} \\ &\quad + h_i\left(\mathbf{v}^n, \nabla \mathbf{v}^n, \frac{\partial \mathbf{v}^n}{\partial t}, \mathbf{T}^n, \nabla \mathbf{T}^n, \mathbf{f}, \nabla \mathbf{f}, \frac{\partial \mathbf{f}}{\partial t}\right) \end{aligned} \tag{4.19}$$

subject to the divergence condition and initial and boundary data

$$\operatorname{div} \mathbf{v}^{n+1} = 0, \ \mathbf{v}^{n+1}(\mathbf{x}, 0) = \mathbf{v}_0(\mathbf{x}), \ \mathbf{v}^{n+1}|_{\partial\Omega} = \mathbf{0}. \tag{4.20}$$

To prove convergence of the iteration, one now has to derive estimates for the solutions of these equations and establish that the mapping which takes $\mathbf{v}^n$ to $\mathbf{v}^{n+1}$ is a contraction in a suitably defined function space. The stress equation (4.17) is a semilinear hyperbolic equation for $\mathbf{T}^n$ which can be dealt with using any of a number of well-known techniques. The velocity equation (4.19), however, is more difficult. Because of the strong ellipticity condition, this equation would be a hyperbolic system were it not for the incompressibility constraint and the presence of the pressure term. The basic idea is the energy method. For the wave equation

$$u_{tt} = \Delta u + f, \tag{4.21}$$

this method involves multiplying the equation by u_t and integrating. If u vanishes on the boundary, the result is

$$\frac{1}{2}\frac{d}{dt}\int_\Omega u_t^2 + |\nabla u|^2\,d\mathbf{x} = \int_\Omega f u_t\,d\mathbf{x}. \tag{4.22}$$

This identity immediately yields a bound for u in terms of f. If we tried to apply the same strategy to (4.19), we would multiply (4.19) by $(\partial/\partial t + (\mathbf{v}\cdot\nabla))v_i$, sum over i, and integrate over Ω. To carry through the nonlinear iteration, we also need to repeat the argument for derivatives, i.e., we have to consider equations resulting from differentiation of (4.19) with respect to time. The technical difficulty that arises in this, however, is that the material time derivative $\partial/\partial t+(\mathbf{v}\cdot\nabla)$ does not take divergence-free vector fields to divergence-free vector fields or gradients to gradients. Overcoming this technical difficulty requires a rather elaborate argument.

There are a number of local time existence results for other models in the literature. Jeffreys-type models, where the stress has an additional Newtonian contribution, are relatively easy, because for local time existence, it is possible to treat the equations as a perturbation of the Newtonian case. In a quite general context, this is carried out in [69]; see also [28]. Integral models are considered in [43], [68], and [70]. In [80], a local existence result for molecular dumbbell models (as discussed in Chapter 2) is established.

If the initial data are small, one expects solutions to exist globally in time and decay to the rest state. The strategy to prove this is to first establish a local existence result and then derive a priori bounds for the solution which ensure the solution remains small in an appropriate norm. There is a very extensive literature on results of this type for various models of viscoelasticity; some of this literature and the basic techniques employed are reviewed in [68]. Most of these results are based on assumptions appropriate to solids rather than fluids, and they do not address the complications resulting from the incompressibility constraint. In [28], a global existence result for differential models of Jeffreys type is established. A model of K-BKZ type was considered in [43]; the technique is extended to general K-BKZ models in [68]. These results are for problems without boundaries; a result for one-dimensional shear flows with boundaries is given in [12]. More general results along such lines could certainly be obtained; doing so would be a laborious technical effort which would, however, not require fundamentally new ideas.

Global existence for large initial data is a much trickier issue. Even for Newtonian fluids, the problem is settled only in two space dimensions and for a few other restricted situations (of course, we are talking only about wall-bounded flows here; we have all observed examples of free surface flows where global existence of smooth solutions fails). For viscoelastic flows, even global existence of weak solutions is an open question; the a priori estimates needed to show this are simply not there, unless we practice "inverse rheology" (i.e., we start with the a priori estimates and work backward to construct the constitutive model). For models of Maxwell type, the governing equations are hyperbolic, and in general we can expect development of shocks in finite time (see Chapter 2 of

[68]). Hence global existence of smooth solutions cannot, in general, be expected for such models, unless a Newtonian contribution is added to the stress. With such an addition, there are a few global existence results in the literature for unidirectional shear flows [27], [51]. There are also some results for integral models [19].

4.2 Steady flows

For steady flows that are small perturbations of the rest state, one can also construct solutions by an iteration which alternates between integrating stresses along streamlines and solving a Stokes-like velocity problem. We shall describe the simplest such result, established in [71]. We consider a differential constitutive law of Maxwell type of the general form

$$(\mathbf{v}\cdot\nabla)\mathbf{T}+\lambda\mathbf{T}+\mathbf{g}(\nabla\mathbf{v},\mathbf{T})=\mu(\nabla\mathbf{v}+(\nabla\mathbf{v})^T). \tag{4.23}$$

Here the function $\mathbf{g}$ is smooth and vanishes at quadratic order:

$$|\mathbf{g}(\nabla\mathbf{v},\mathbf{T})|\le C(|\nabla\mathbf{v}|^2+|\mathbf{T}|^2) \tag{4.24}$$

for small values of $\nabla\mathbf{v}$ and $\mathbf{T}$. We have to solve this constitutive relation in conjunction with the equation of motion

$$\rho(\mathbf{v}\cdot\nabla)\mathbf{v}=\operatorname{div}\mathbf{T}-\nabla p+\mathbf{f}, \tag{4.25}$$

the incompressibility condition

$$\operatorname{div}\mathbf{v}=0, \tag{4.26}$$

and Dirichlet boundary conditions for the velocity

$$\mathbf{v}|_{\partial\Omega}=\mathbf{v}_0, \tag{4.27}$$

where $\mathbf{v}_0$ is tangent to $\partial\Omega$.

The following result is established in [71].

Theorem 2 *Assume that Ω is a bounded smooth domain and that $\mathbf{f}\in H^2(\Omega)$, $\mathbf{v}_0\in H^{5/2}(\partial\Omega)$ with $\|\mathbf{f}\|_2$ and $\|\mathbf{v}_0\|_{5/2}$ sufficiently small. Then there exists a solution with the regularity $\mathbf{v}\in H^3(\Omega)$, $\mathbf{T}\in H^2(\Omega)$, and $p\in H^2(\Omega)$. This solution is unique among small solutions.*

To set up an iterative construction, we proceed in a similar fashion as we did above for initial value problems. We apply the divergence operator to the constitutive relation and find

$$(\mathbf{v}\cdot\nabla)\operatorname{div}\mathbf{T}+\lambda\operatorname{div}\mathbf{T}+\mathbf{h}(\nabla\mathbf{v},\nabla^2\mathbf{v},\mathbf{T},\nabla\mathbf{T})=\mu\Delta\mathbf{v}, \tag{4.28}$$

where $\mathbf{h}$ is a nonlinear expression that depends on $\mathbf{T}$ and its first derivatives as well as first and second derivatives of $\mathbf{v}$. The crucial fact is that $\mathbf{h}$ vanishes at quadratic order if its arguments tend to zero. We then substitute div **T** from the momentum equation into (4.28). The result is an equation of the form

$$\mu\Delta\mathbf{v} - \nabla[(\mathbf{v}\cdot\nabla)p + \lambda p] = -(\mathbf{v}\cdot\nabla)\mathbf{f} - \lambda\mathbf{f} + \mathbf{r}(\mathbf{v}, \nabla\mathbf{v}, \nabla^2\mathbf{v}, \mathbf{T}, \nabla\mathbf{T}, \nabla p). \quad (4.29)$$

Here **r** is a function which vanishes quadratically when its arguments tend to zero.

The iteration can now be described as follows. We set $q = (\mathbf{v}\cdot\nabla)p + \lambda p$. Given an iterate $(\mathbf{v}^n, q^n)$, we determine p^n and $\mathbf{T}^n$ by

$$\begin{aligned} (\mathbf{v}^n\cdot\nabla)p^n + \lambda p^n &= q^n, \\ (\mathbf{v}^n\cdot\nabla)\mathbf{T}^n + \lambda\mathbf{T}^n + \mathbf{g}(\nabla\mathbf{v}^n, \mathbf{T}^n) &= \mu(\nabla\mathbf{v}^n + (\nabla\mathbf{v}^n)^T). \end{aligned} \quad (4.30)$$

Then we calculate the new iterate for **v** and q from

$$\mu\Delta\mathbf{v}^{n+1} - \nabla q^{n+1} = -(\mathbf{v}^n\cdot\nabla)\mathbf{f} - \lambda\mathbf{f} + \mathbf{r}(\mathbf{v}^n, \nabla\mathbf{v}^n, \nabla^2\mathbf{v}^n, \mathbf{T}^n, \nabla\mathbf{T}^n, \nabla p^n) \quad (4.31)$$

subject to the incompressibility condition

$$\operatorname{div}\mathbf{v}^{n+1} = 0 \quad (4.32)$$

and the boundary condition

$$\mathbf{v}^{n+1}|_{\partial\Omega} = \mathbf{v}_0. \quad (4.33)$$

At each step of the iteration, we have to solve a semilinear hyperbolic equation to determine the stress and pressure and then a Stokes problem to find a new velocity and a new q. The convergence of the iteration is established by combining estimates for these two types of problem.

The result can be generalized in a number of ways. Already in [71], it has been pointed out how to extend the procedure to models with several relaxation modes and to Jeffreys-type models. A generalization to integral models is given in [73]. Guillopé and Saut [29] consider the situation where the state being perturbed is not the rest state, but a given Newtonian flow; naturally, the assumption needs to be made that non-Newtonian effects are weak. The approach has also been extended to exterior flows [55].

For Newtonian fluids, it makes little difference whether the velocity is required to satisfy homogeneous or inhomogeneous boundary conditions. For non-Newtonian fluids, this is so only as long as the velocity on the boundary is tangent to the boundary, as we assumed above. This is because the fluid has memory, and therefore, if the fluid enters through the boundary of the domain, then some information about the prior flow history needs to be provided to have a well-posed problem. The precise nature of this information is model dependent, and only differential models have been investigated in this context. If we try to apply the iteration above in a situation where the domain has inflow and outflow boundaries, then two things change. First, we need to know the value of $\mathbf{T}^n$ and p^n at the inflow boundary to solve (4.30). Second, there is the issue of getting back from (4.29) to the original momentum equation. Recall that (4.29) was derived by applying $(\mathbf{v}\cdot\nabla) + \lambda$ to the momentum equation and combining the resulting equation with the divergence of the constitutive equation. Hence if (4.29) and the constitutive equation are satisfied, then it follows that

$$[(\mathbf{v}\cdot\nabla)+\lambda](\text{momentum equation})=0. \tag{4.34}$$

If there are no inflow boundaries, this is no problem, but if there are inflow boundaries, then we can get back from (4.34) to the momentum equation only if the momentum equation

$$\rho(\mathbf{v}\cdot\nabla)\mathbf{v}=\operatorname{div}\mathbf{T}-\nabla p+\mathbf{f} \tag{4.35}$$

is satisfied at the inflow boundary. This condition is an implicit constraint on the boundary data which can be prescribed at the inflow boundary. This makes the characterization of inflow boundary data a nontrivial problem. The idea is as follows. Suppose for simplicity that the inflow boundary is a plane $x=\text{const}$. The equation (4.35) has the form

$$\begin{aligned}
\frac{\partial T_{11}}{\partial x}+\frac{\partial T_{12}}{\partial y}+\frac{\partial T_{13}}{\partial z}-\frac{\partial p}{\partial x} &= \cdots,\\
\frac{\partial T_{12}}{\partial x}+\frac{\partial T_{22}}{\partial y}+\frac{\partial T_{23}}{\partial z}-\frac{\partial p}{\partial y} &= \cdots,\\
\frac{\partial T_{13}}{\partial x}+\frac{\partial T_{23}}{\partial y}+\frac{\partial T_{33}}{\partial z}-\frac{\partial p}{\partial z} &= \cdots,
\end{aligned} \tag{4.36}$$

where the dots indicate terms involving the given body force and the velocity, which we can consider known at the stage of the iteration where a new stress needs to be constructed. Next we can use the equation $(\mathbf{v}\cdot\nabla)p+\lambda p=q$ to eliminate $\partial p/\partial x$: if $\mathbf{v}=(u,v,w)$, we have

$$\frac{\partial p}{\partial x}=-\frac{v}{u}\frac{\partial p}{\partial y}-\frac{w}{u}\frac{\partial p}{\partial z}-\frac{\lambda}{u}p+\frac{q}{u}. \tag{4.37}$$

Similarly, we can use the constitutive law to eliminate x-derivatives of the stresses. Inserting the result into (4.36), we end up with a system involving only the values of the stresses and their derivatives tangent to the boundary. We now need to use this system to eliminate some of the stress components and determine their values in terms of the others. At each step of the iteration, we must then, prior to integrating the stresses, solve this problem on the inflow boundary.

If one is looking for a simple situation to consider inflow boundaries, it is natural to study flows which perturb a uniform flow or a parallel shear flow; some existence results for such flows and characterizations of proper boundary conditions were obtained in [72, 74, 75, 77, 79, 81, 88]. The problem of inflow boundaries for time-dependent flows is even more difficult; only a very special case has been solved [91].

Chapter 5

Numerical Methods

Efforts in numerical simulation of polymeric flows began in the late 1970s and had their first successes in the mid 1980s. Since then, the importance of numerical simulations in rheology has steadily increased. The success of numerical simulations is largely responsible for a shift in interest toward complex flows.

A comprehensive discussion of numerical methods is beyond the scope of these lecture notes. Rather, we shall confine ourselves to a brief outline of the main issues, with special attention to points which are specific to the equations governing viscoelastic flows. Crochet's review article [13] is a good starting point for further reading. For the discussion in this section, we shall confine ourselves to steady flows of fluids described by Maxwell-type differential models.

In numerical simulation, there are basically four issues to be addressed:

1. On which equations should the numerical simulation be based?

2. Which type of discrete representation should be used to represent the solution?

3. What is the discrete representation of the equations?

4. How does one go about solving the nonlinear system of discretized equations?

For each of these points, there are a number of choices available, which has led to a rich proliferation of concatenated acronyms. The first question might seem a stupid one, since the equations are given. However, we already found it advantageous in the chapter on existence theory (Chapter 4) to recombine the equations in a different form, and such recombinations are also important for numerical simulation. Indeed, it is basically this issue and the issue of streamline upwinding discussed later that are the crucial points in numerical simulation of viscoelastic flows. Before discussing these issues in a little more detail, we shall make some brief remarks about the other points.

5.1 Discrete representation

There are three basic approaches to discretize a partial differential equation:

1. *Finite difference methods* represent the solution by its values on a grid. The derivatives occurring in the equations are then approximated by difference quotients computed from the values on the grid. This method is conceptually the simplest, but it becomes awkward to use if complex geometries are involved.

2. *Finite element methods* divide the flow domain into triangles or quadrilaterals and represent the solution as a linear superposition of piecewise polynomials defined on these subregions. The equations are interpreted in a "weak" sense, i.e., the discretized equations are obtained by integrating the original equation against a "test function" which is, in the simplest case, chosen from the same space as that used to represent the solution. This method is the most versatile in representing complex geometries.

3. *Spectral methods* represent the solution in terms of orthogonal polynomials or trigonometric functions. The equations are then expanded in the same fashion and discretized either by truncation or by "collocation," i.e., evaluation at optimally chosen points. The great advantage of these methods is their high accuracy.

There are also combined methods, e.g., for a flow domain that is a union of several rectangles, we may divide the domain into rectangles (finite elements) and then use a spectral discretization within each rectangle.

All the approaches outlined above have been used in the simulation of viscoelastic flows.

5.2 Nonlinear iteration

At the end of the discretization process, we have a set of nonlinear equations, which we may put in the schematic form

$$F(q) = 0. \tag{5.1}$$

Here both q and F take values in some finite-dimensional space $\mathbb{R}^m$.

One of the principal limitations of mortal beings is that they do not know how to solve nonlinear systems, and hence we need to replace the nonlinear system with a sequence of linear ones. The main approaches to this are Newton iteration and Picard iteration. Newton iteration proceeds by linearizing the equation at each iterate:

$$F(q_n) + DF(q_n)(q_{n+1} - q_n) = 0. \tag{5.2}$$

Here DF denotes the Jacobian matrix with entries $\partial F_i/\partial q_j$. The advantage of Newton's iteration is that it is guaranteed to converge rapidly if the starting value is close enough to the solution; the disadvantage is that it requires the evaluation and inversion of the Jacobian.

Picard iteration is based on representing the system in the form

$$F(q) = G(q)q + H(q) = 0, \tag{5.3}$$

where $G(q)$ is a matrix, and then solving iteratively in the following fashion:

$$G(q_n)q_{n+1} + H(q_n) = 0. \tag{5.4}$$

This is usually simpler to implement than Newton iteration; the drawback is that convergence is guaranteed only in restricted situations (e.g., if the flow is close enough to the rest state). Indeed, the iteration we used in the discussion of existence in the previous chapter can be viewed as a form of Picard iteration.

Regardless of which type of iteration is used, it needs a starting value, and typically there is no good reason to hope for convergence unless that starting value is already close to the solution. One way to address this is by using a continuation method. For instance, we can start by solving for Stokes flow, which is a linear problem, and then gradually increase the Weissenberg number in small steps, each time using the solution from the previous Weissenberg number as a starting point for the iteration at the new Weissenberg number.

5.3 Choice of equations

The simplest choice of equations is, of course, the one in which the equations are usually given. That is, we have the momentum equation, the incompressibility condition, and a constitutive relation:

$$\begin{aligned} \operatorname{div} \mathbf{T} - \nabla p - \rho(\mathbf{v} \cdot \nabla)\mathbf{v} &= \mathbf{0}, \\ \operatorname{div} \mathbf{v} &= 0, \\ (\mathbf{v} \cdot \nabla)\mathbf{T} + \lambda \mathbf{T} + \mathbf{g}(\mathbf{T}, \nabla \mathbf{v}) - \mu(\nabla \mathbf{v} + (\nabla \mathbf{v})^T) &= \mathbf{0}. \end{aligned} \tag{5.5}$$

If we use a finite-element discretization, we approximate $\mathbf{v}$ by an element of a finite-dimensional space V_h, p by an element of a finite-dimensional space P_h, and $\mathbf{T}$ by an element of a finite-dimensional space T_h. The equations are represented by the finite system

$$\begin{aligned} \int_\Omega (\mathbf{T} - p\mathbf{I}) : \nabla \mathbf{u} + \rho[(\mathbf{v} \cdot \nabla)\mathbf{v}] \cdot \mathbf{u} \, d\mathbf{x} &= 0, \\ \int_\Omega (\operatorname{div} \mathbf{v}) q \, d\mathbf{x} &= 0, \\ \int_\Omega \left[(\mathbf{v} \cdot \nabla)\mathbf{T} + \lambda \mathbf{T} + \mathbf{g}(\mathbf{T}, \nabla \mathbf{v}) - \mu(\nabla \mathbf{v} + (\nabla \mathbf{v})^T)\right] : \mathbf{S} \, d\mathbf{x} &= 0 \end{aligned} \tag{5.6}$$

for every $\mathbf{u} \in V_h$, $q \in P_h$, $\mathbf{S} \in T_h$. Here Ω is the flow domain, and we have, for simplicity, assumed homogeneous Dirichlet conditions for the velocity.

The naive implementation of this or similar approaches in the early 1980s led to universal failure of numerical simulations, which became referred to as the "high Weissenberg number problem," although actually "nonzero Weissenberg number problem" is the more appropriate term. One of the problems is that a straightforward Galerkin discretization of the constitutive law has poor stability properties if the advection term $(\mathbf{v} \cdot \nabla)\mathbf{T}$ becomes dominant. We shall discuss approaches to deal with this issue in section 5.4. The other problem is that the equations for a viscoelastic fluid are of a combined elliptic-hyperbolic type, and the behavior of such equations under discretization is poorly understood. Marchal and Crochet [54] have shown that the "mixed" finite-element method as described above will actually work if an appropriate streamline upwinding is used, and, in addition, the spaces for velocities, stresses, and pressure are chosen appropriately. In addition to the usual Babuška–Brezzi (inf-sup) condition for velocity and pressure (see, e.g., [22]), it is necessary to use a combination of elements that resolves the stresses with higher accuracy than the velocities.

Another approach, however, is to reformulate the mixed system in a way that tries to separate the elliptic part from the hyperbolic part. Indeed, this is what we did in the discussion of existence theory in Chapter 4. One possibility is to use a system of equations consisting of (4.29), the incompressibility condition, and the constitutive equation. We can then, as we did in Chapter 4, regard (4.29) and the incompressibility condition as an elliptic Stokes-like problem for velocity and pressure, and the constitutive law as a hyperbolic equation for determining the stress. This approach has become known as the explicitly elliptic momentum equation (EEME) method. It was introduced to numerical simulations by King et al. [44]. The EEME method has been quite successful in numerical simulations of viscoelastic flows. Its drawbacks are that it becomes much harder to implement if multimode or Jeffreys-type models are to be considered, and it also does not have a natural way of implementing free surface boundary conditions.

Another, much simpler way of modifying the equations is known as the elastic-viscous stress-splitting (EVSS) method. In this method, we make the substitution

$$\mathbf{T} = \frac{\mu}{\lambda}(\nabla\mathbf{v} + (\nabla\mathbf{v})^T) + \mathbf{S}. \tag{5.7}$$

When we insert this into (5.5), we obtain

$$\begin{aligned}
\operatorname{div}\mathbf{S} + \frac{\mu}{\lambda}\Delta\mathbf{v} - \nabla p - \rho(\mathbf{v}\cdot\nabla)\mathbf{v} &= \mathbf{0},\\
\operatorname{div}\mathbf{v} &= 0,\\
(\mathbf{v}\cdot\nabla)\left[\mathbf{S} + \frac{\mu}{\lambda}(\nabla\mathbf{v} + (\nabla\mathbf{v})^T)\right]&\\
+\lambda\mathbf{S} + \mathbf{g}\left(\mathbf{S} + \frac{\mu}{\lambda}(\nabla\mathbf{v} + (\nabla\mathbf{v})^T), \nabla\mathbf{v}\right) &= \mathbf{0}.
\end{aligned} \tag{5.8}$$

Note that this introduces the "elliptic" term $\frac{\mu}{\lambda}\Delta\mathbf{v}$ into the momentum equation; indeed in the Newtonian case, $\mathbf{S} = \mathbf{0}$, and the momentum equation assumes the usual Navier–Stokes form. The price we pay, however, is the appearance of the convected derivative of the velocity gradient in the constitutive equation.

This equation now contains second derivatives of the velocity rather than just first derivatives. As pointed out in [66], where the EVSS formulation was first applied successfully, earlier attempts failed because of the way they handled the discretization of this term.

The discrete EVSS (DEVSS) method [26] has the stabilizing feature of the EVSS method in the momentum equation, but avoids the problem of introducing a convected derivative of the velocity gradient into the constitutive equation. In the DEVSS method, we write the momentum equation in the form

$$\operatorname{div}\mathbf{S} + \frac{\mu}{\lambda}\operatorname{div}(\nabla\mathbf{v} + (\nabla\mathbf{v})^T) - \frac{\mu}{\lambda}\operatorname{div}\mathbf{D} - \nabla p - \rho(\mathbf{v}\cdot\nabla)\mathbf{v} = \mathbf{0}.$$

Nothing is changed in the constitutive equation. The variable $\mathbf{D}$ is equal to $\nabla\mathbf{v} + (\nabla\mathbf{v})^T$ at the continuous level, so we really have not changed the equation at all. At the discrete level, however, $\mathbf{D}$ is discretized in the same fashion as the stress $\mathbf{S}$, so if the stresses are approximated with an accuracy which does not fully capture $\nabla\mathbf{v} + (\nabla\mathbf{v})^T$, then $\mathbf{D}$ is a projection of $\nabla\mathbf{v} + (\nabla\mathbf{v})^T$. Doing this allows flexibility in choosing elements for velocities and stresses while maintaining numerical stability.

5.4 Streamline upwinding

As we already mentioned, a straightforward Galerkin discretization of the constitutive equations leads to poor results when the advection term $(\mathbf{v}\cdot\nabla)\mathbf{T}$ becomes dominant. A number of approaches can be used to improve the stability of the numerical method in this case. One approach is artificial diffusion. Here one adds a term $h\Delta\mathbf{T}$ to the constitutive equation, where h is proportional to the mesh size. Alternatively, one can consider anisotropic diffusion terms where diffusion acts only in the direction of the streamlines. While such diffusion terms improve stability, they also lead to a loss of accuracy, because an extraneous term has been added to the equation.

Upwinding methods, in the context of finite-difference methods, mean the replacement of centered differences with asymmetric ones. In the finite-element method, upwinding is a change in the test function. That is, instead of discretizing the constitutive equation as in (5.6), we use the following discretization:

$$\begin{aligned}\int_\Omega &\big[(\mathbf{v}\cdot\nabla)\mathbf{T} + \lambda\mathbf{T} + \mathbf{g}(\mathbf{T},\nabla\mathbf{v}) \\ &-\mu(\nabla\mathbf{v} + (\nabla\mathbf{v})^T)\big] : (\mathbf{S} + h(\mathbf{v}\cdot\nabla)\mathbf{S})\,d\mathbf{x} = 0. \qquad (5.9)\end{aligned}$$

That is, we actually have not changed the equation itself, but the test function which is used for the discrete implementation of the equation. In contrast to the usual Galerkin method, the test functions are not the same as those used to expand the solution. A thorough discussion of upwinding methods for finite-element solutions of hyperbolic problems is given in [39].

The numerical simulation of viscoelastic flows at low to moderate Weissenberg numbers and in smooth geometries is, from a practical point of view, a solved problem at this point. Many different methods have been used successfully. Little is known overall in terms of rigorous convergence theory, but a few studies have appeared over the past few years [1, 2, 3]. The flows covered by these studies are small perturbations of the rest state, as in the existence theories discussed in the section 5.3. The main practical problems that remain in numerical simulations of viscoelastic flows concern features of the high Weissenberg number limit, where steep stress gradients form along boundaries or separating streamlines. There are also difficulties in problems with stress singularities. This is a new high Weissenberg number problem of a quite different nature than that of the early 1980s. In the next chapter, we shall discuss the current state of efforts to gain a mathematical understanding of the high Weissenberg number behavior of viscoelastic flows.

Chapter 6

High Weissenberg Number Asymptotics

The flow of Newtonian fluids at high Reynolds number presents a host of mathematical difficulties. Formally, if we take the Reynolds number to infinity, we obtain the Euler equations. Unfortunately, however, the Euler equations often tell us little about the actual flow behavior, because solutions are highly nonunique. For instance, if we consider flow in a pipe, then every parallel velocity profile satisfies the Euler equations, and we learn absolutely nothing about the flow which is actually established. Another difficulty of high Reynolds number flow is the formation of singular layers along boundaries and separating streamlines, where the validity of the Euler equations breaks down. Finally, there is the question of instabilities and complex dynamics.

It turns out that all of these difficulties exist also for the high Weissenberg number limit of non-Newtonian flows. Indeed, the most severe limitations on successful numerical simulations at this time are linked to the difficulty of resolving high-stress gradients arising in boundary layers, along separating streamlines, and near corner singularities. The problems are most severe when the UCM model is chosen as a constitutive theory. This should be no surprise: As we saw in the chapter on simple flows (Chapter 3), the UCM model predicts higher stresses at large deformation rates than other popular models, and so we may expect high Weissenberg number features to be more pronounced.

In this chapter, we review some recent progress toward understanding the nature of high Weissenberg number flows using asymptotic methods. We shall confine our discussion to two-dimensional steady flows, and we shall mostly be concerned with the UCM model. The objective is a formal asymptotic description of a flow which we presume to exist; we do not have the means of establishing existence of steady flows at arbitrary Weissenberg number, and we certainly cannot claim that such flows will remain stable.

6.1 Euler equations

We consider the steady, two-dimensional creeping flow of a UCM fluid, described by the following dimensionless equations:

$$\begin{aligned} \operatorname{div} \mathbf{T} - \nabla p &= \mathbf{0}, \\ \operatorname{div} \mathbf{v} &= 0, \\ (\mathbf{v} \cdot \nabla)\mathbf{T} - (\nabla \mathbf{v})\mathbf{T} - \mathbf{T}(\nabla \mathbf{v})^T + W^{-1}\mathbf{T} &= W^{-1}(\nabla \mathbf{v} + (\nabla \mathbf{v})^T). \end{aligned} \tag{6.1}$$

We are interested in studying the asymptotic behavior of solutions in the limit $W \to \infty$. Clearly, the most naive thing to do is simply set $W = \infty$, which changes the last equation of (6.1) to

$$(\mathbf{v} \cdot \nabla)\mathbf{T} - (\nabla \mathbf{v})\mathbf{T} - \mathbf{T}(\nabla \mathbf{v})^T = \mathbf{0}. \tag{6.2}$$

As a first step of studying the high Weissenberg number asymptotics for the UCM fluid, it is therefore of interest to look at (6.2) together with the momentum and incompressibility equations. These equations have the same status for high Weissenberg number flows that the Euler equations have for high Reynolds number flows. As we shall see, there is actually a connection with the Euler equations. The analysis of this section follows [92].

To analyze (6.2), it is convenient to transform the stress tensor to a different basis. For this purpose, we introduce a vector $\mathbf{w}$ that is perpendicular to $\mathbf{v}$ and such that $\|\mathbf{v} \times \mathbf{w}\| = 1$:

$$\mathbf{w} = \left(-\frac{v_2}{v_1^2 + v_2^2}, \frac{v_1}{v_1^2 + v_2^2} \right). \tag{6.3}$$

We represent the stress tensor in a basis spanned by the vectors $\mathbf{v}$ and $\mathbf{w}$:

$$\mathbf{T} = \lambda \mathbf{v}\mathbf{v}^T + \mu(\mathbf{v}\mathbf{w}^T + \mathbf{w}\mathbf{v}^T) + \nu \mathbf{w}\mathbf{w}^T. \tag{6.4}$$

This transforms (6.2) to the following set of equations:

$$\begin{aligned} (\mathbf{v} \cdot \nabla)\lambda + 2\mu \operatorname{div} \mathbf{w} &= 0, \\ (\mathbf{v} \cdot \nabla)\mu + \nu \operatorname{div} \mathbf{w} &= 0, \\ (\mathbf{v} \cdot \nabla)\nu &= 0. \end{aligned} \tag{6.5}$$

We further note that we can expect $\mathbf{T}$ to be approximated by a rank-one tensor in the high Weissenberg number limit. To see this, we note that it follows from (6.1) that

$$(\mathbf{v} \cdot \nabla) \left(\det \left(\mathbf{T} + \frac{1}{W} \right) \right) = -\frac{1}{W} \det \left(\mathbf{T} + \frac{1}{W} \right) \operatorname{tr} \left(\left(\mathbf{T} + \frac{1}{W} \right)^{-1} \mathbf{T} \right). \tag{6.6}$$

If both eigenvalues of $\mathbf{T}$ are positive, the right-hand side of (6.6) is negative, so the determinant of $\mathbf{T} + 1/W$ decreases along streamlines. Unless one imposes unphysical upstream conditions, $\mathbf{T}$ cannot have two eigenvalues of order 1. For a rank-one tensor, we can set $\lambda = \alpha^2$, $\mu = \alpha\beta$, $\nu = \beta^2$ in (6.4). We find then that (6.5) reduces to the two equations

$$\begin{aligned} (\mathbf{v} \cdot \nabla)\alpha + \beta \operatorname{div} \mathbf{w} &= 0, \\ (\mathbf{v} \cdot \nabla)\beta &= 0. \end{aligned} \tag{6.7}$$

We introduce the following new quantities:

$$\mathbf{u} = K^{-1}(\alpha\beta\mathbf{v} + \beta^2\mathbf{w}), \ \rho = K^2\beta^{-2}, \tag{6.8}$$

where K is an arbitrary constant. (By introducing the constant K, we can formally accommodate the limiting case $\beta = 0$ by letting K tend to zero in such a way that $K/\beta = 1$.) We now find

$$\operatorname{div}(\rho\mathbf{u}) = K \operatorname{div}\left(\frac{\alpha}{\beta}\mathbf{v} + \mathbf{w}\right) = \frac{K}{\beta}(\mathbf{v} \cdot \nabla)\alpha + K \operatorname{div} \mathbf{w} = 0, \tag{6.9}$$

where we have taken account of (6.7) and the incompressibility condition $\operatorname{div} \mathbf{v} = 0$. Moreover, we have $\mathbf{T} = \rho\mathbf{u}\mathbf{u}^T$ according to (6.4). Inserting this into the momentum equation, we find

$$\rho(\mathbf{u} \cdot \nabla)\mathbf{u} = \nabla p. \tag{6.10}$$

We note that, apart from a sign change in the pressure, (6.9) and (6.10) are precisely the Euler equations. Of course, the interpretation of variables is different, since $\mathbf{u}$ does not represent the velocity and ρ does not represent the density. Moreover, there is no equation of state linking p and ρ. We next consider the question of what replaces the equation of state.

To get back from ρ and $\mathbf{u}$ to the original variables, we must set

$$\mathbf{u} = \alpha\rho^{-1/2}\mathbf{v} + K\rho^{-1}\mathbf{w}, \ \mathbf{T} = \rho\mathbf{u}\mathbf{u}^T, \tag{6.11}$$

and we must satisfy

$$\operatorname{div} \mathbf{v} = 0, \ (\mathbf{v} \cdot \nabla)\rho = 0, \ \mathbf{v} \cdot \mathbf{w} = 0, \ \mathbf{v} \times \mathbf{w} = \mathbf{e}_3. \tag{6.12}$$

We can eliminate $\mathbf{w}$ by taking the vector product of the first equation of (6.11) with $\mathbf{v}$. This results in

$$\mathbf{v} \times (\rho\mathbf{u}) = K\mathbf{e}_3, \ \mathbf{v} \cdot \nabla\rho = 0, \ \operatorname{div} \mathbf{v} = 0. \tag{6.13}$$

If $\mathbf{u}$ and $\nabla\rho$ are not orthogonal, then the first two equations of (6.13) determine $\mathbf{v}$ uniquely, and the last equation can therefore be viewed as a constraint on $\mathbf{u}$ and ρ. To make this constraint explicit, we take the vector product of the first equation in (6.13) with $\nabla\rho$. This yields

$$\begin{aligned} \nabla\rho \times (\mathbf{v} \times (\rho\mathbf{u})) &= \rho(\mathbf{u} \cdot \nabla\rho)\mathbf{v} - (\mathbf{v} \cdot \nabla\rho)(\rho\mathbf{u}) = -(\rho^2 \operatorname{div} \mathbf{u})\mathbf{v} \\ = K\nabla\rho \times \mathbf{e}_3 &= K\nabla \times (\rho\mathbf{e}_3). \end{aligned} \tag{6.14}$$

Taking the divergence on both sides, we find

$$-\rho^2 \operatorname{div} \mathbf{u} \operatorname{div} \mathbf{v} - \rho^2(\mathbf{v} \cdot \nabla) \operatorname{div} \mathbf{u} = 0. \tag{6.15}$$

Hence the condition $\operatorname{div} \mathbf{v} = 0$ is equivalent to $(\mathbf{v} \cdot \nabla)(\operatorname{div} \mathbf{u}) = 0$, which is equivalent to saying that $\nabla \operatorname{div} \mathbf{u}$ is in the same direction as $\nabla\rho$, i.e., $\operatorname{div} \mathbf{u}$ is a function of ρ.

There are also, on the other hand, solutions of (6.9) and (6.10) for which $\mathbf{u} \cdot \nabla\rho$ vanishes identically. In that case, (6.9) yields $\operatorname{div} \mathbf{u} = 0$. The equations (6.13) are then consistent only if either $K = 0$ or $\nabla\rho = \mathbf{0}$. In the first case, $\mathbf{u}$ is parallel to $\mathbf{v}$ and $\beta = 0$. In the second case, the second equation of (6.13) is vacuous and we can use the first equation to determine $\mathbf{v}$ up to a multiple of $\mathbf{u}$: $\mathbf{v} = \mathbf{v}_0 + \lambda\mathbf{u}$. The condition $\operatorname{div} \mathbf{v} = 0$ then yields a first-order differential equation for λ.

In summary, we have found a complete set of equations given by

$$\begin{aligned} \rho(\mathbf{u} \cdot \nabla)\mathbf{u} &= \nabla p, \\ \operatorname{div}(\rho\mathbf{u}) &= 0, \\ \operatorname{div} \mathbf{u} &= \phi(\rho), \end{aligned} \tag{6.16}$$

where ϕ is an arbitrary function. Any solution of these equations yields a corresponding solution for the infinite Weissenberg number limit of (6.1). Of course, there is a high degree of nonuniqueness here; even more than for the Euler equations, since we have the arbitrary function ϕ. In many flow situations, however, upstream conditions are such that $\mathbf{u}$ is parallel to $\mathbf{v}$ and $\phi(\rho)$ is zero. In this case, we simply recover the incompressible Euler equations.

An important class of solutions of the Euler equations is generated by potential flow. In (6.16), we have, in general, no relationship between p and ρ, so, in contrast to inviscid fluid mechanics, we cannot claim that $\nabla p/\rho$ is a gradient. Nevertheless, if we specialize the choice of the function $\phi(\rho)$, we can find an analogue of potential flow. We write

$$\begin{aligned} \rho(\mathbf{u} \cdot \nabla)\mathbf{u} &= [(\rho^{1/2}\mathbf{u}) \cdot \nabla](\rho^{1/2}\mathbf{u}) - \frac{1}{2}[(\mathbf{u} \cdot \nabla)\rho]\mathbf{u} \\ &= [(\rho^{1/2}\mathbf{u}) \cdot \nabla](\rho^{1/2}\mathbf{u}) + \frac{1}{2}(\rho \operatorname{div} \mathbf{u})\mathbf{u}. \end{aligned} \tag{6.17}$$

According to (6.16), $\operatorname{div} \mathbf{u}$ should be a function of ρ, and we now make the special choice

$$\operatorname{div} \mathbf{u} = A\rho^{-1/2} \tag{6.18}$$

with a constant A. With this choice, we can satisfy the momentum equation if we set

$$\rho^{1/2}\mathbf{u} = \nabla q, \ p = \frac{1}{2}|\nabla q|^2 + \frac{1}{2}Aq. \tag{6.19}$$

With this choice, (6.18) assumes the form

$$\operatorname{div}(\rho^{-1/2}\nabla q) = A\rho^{-1/2}, \tag{6.20}$$

Figure 6.1: Flow past a cylinder in a channel. Reprinted from [59], P.J. Oliveira, F.T. Pinhoi, and G.A. Pinto, Numerical simulation of non-linear elastic flows with a general collocated finite-volume method, *J. Non-Newt. Fluid Mech.*, 79 (1998), 1–43, with permission from Elsevier Science.

and the second equation in (6.16) becomes

$$\operatorname{div}(\rho^{1/2}\nabla q) = 0. \tag{6.21}$$

We can eliminate ρ between these two equations to obtain

$$\Delta q = \frac{A}{2}. \tag{6.22}$$

6.2 Boundary layers

The validity of the analysis above breaks down near solid boundaries and separating streamlines, where singular layers form. An illustration of such features can be seen in Figure 6.1. This figure is from [59]. The problem under study is flow past a cylinder constrained by a channel, and the figure shows half the flow domain. The flow is from left to right, and the plot shows contour lines of the stress difference $T_{xx} - T_{yy}$, where the x-axis is pointing to the right and the y-axis is pointing up. We can clearly see the concentrations of stress gradients along the edge of the cylinder, along the wall of the channel, and in the wake of the cylinder. Analogous features are found in many other flow geometries, e.g., flow between eccentric cylinders, flow in an undulating tube, and driven cavity flow.

Although these singular layers are somewhat analogous to high Reynolds number boundary layers in Newtonian flows, the reason for their appearance is a quite different one. Boundary layers in Newtonian flows appear because solutions of the Euler equations generally do not satisfy the no-slip boundary condition for the velocity. In contrast, high Weissenberg number boundary layers are features of the stress, not the velocity; they appear even if the stresses are integrated in a prescribed velocity field. The reason for the boundary layers is the memory of the fluid. At the boundary, particles do not move, and hence the stress is completely determined by the local shear rate. On the other hand, at a small distance d from the boundary, particles cover a distance of order Wd within one relaxation time, and hence the memory of the fluid is important if W is large. This leads to a sharp transition in the stress. Along separating streamlines, as in the wake of the cylinder, the stresses are strongly influenced by the stretching flow which occurs near the stagnation point where the streamline separates from the wall.

We shall now discuss the derivation of boundary layer equations for the UCM fluid (see [93]). We begin with the equations for creeping flow of a UCM fluid in two space dimensions, which we state in dimensionless form:

$$\begin{aligned}
W\left[u\frac{\partial T_{11}}{\partial x}+v\frac{\partial T_{11}}{\partial y}-2\frac{\partial u}{\partial x}T_{11}-2\frac{\partial u}{\partial y}T_{12}\right]+T_{11} &= 2\frac{\partial u}{\partial x},\\
W\left[u\frac{\partial T_{12}}{\partial x}+v\frac{\partial T_{12}}{\partial y}-\frac{\partial u}{\partial y}T_{22}-\frac{\partial v}{\partial x}T_{11}\right]+T_{12} &= \frac{\partial u}{\partial y}+\frac{\partial v}{\partial x},\\
W\left[u\frac{\partial T_{22}}{\partial x}+v\frac{\partial T_{22}}{\partial y}-2\frac{\partial v}{\partial x}T_{12}-2\frac{\partial v}{\partial y}T_{22}\right]+T_{22} &= 2\frac{\partial v}{\partial y},\\
\frac{\partial u}{\partial x}+\frac{\partial v}{\partial y} &= 0,\\
\frac{\partial^2}{\partial x\partial y}(T_{11}-T_{22})+\left(\frac{\partial^2}{\partial y^2}-\frac{\partial^2}{\partial x^2}\right)T_{12} &= 0.
\end{aligned} \tag{6.23}$$

For simplicity, let us assume a flat boundary. Thus the flow domain is the half-plane $y>0$, and we are interested in solutions for small y and large W. A self-consistent scaling of variables can be motivated as follows: At the wall $y=0$, the velocity vanishes and the stresses are viscometric. If the shear rate is of order 1, then the shear stress T_{12} is of order 1 and the first normal stress T_{11} is of order W. In order to balance $\partial T_{11}/\partial x$ with $\partial T_{12}/\partial y$ in the momentum equation, we should therefore scale y with a factor $1/W$. If the shear rate is of order 1, then at a distance of order $1/W$ from the wall, the velocity is also of order $1/W$. In this fashion, we obtain the following self-consistent scaling of variables:

$$\begin{gathered}
y=z/W,\ u=\tilde{u}(x,z)/W,\ v=\tilde{v}(x,z)/W^2,\\
T_{11}=W\tilde{T}_{11}(x,z),\ T_{12}=\tilde{T}_{12}(x,z),\ T_{22}=\tilde{T}_{22}(x,z)/W.
\end{gathered} \tag{6.24}$$

We insert these scalings into the governing equations, keep only the leading-order terms in W, and suppress the tildes. This leads to the following set of boundary layer equations:

$$\begin{aligned}
u\frac{\partial T_{11}}{\partial x}+v\frac{\partial T_{11}}{\partial z}-2\frac{\partial u}{\partial x}T_{11}-2\frac{\partial u}{\partial z}T_{12}+T_{11} &= 0,\\
u\frac{\partial T_{12}}{\partial x}+v\frac{\partial T_{12}}{\partial z}-\frac{\partial u}{\partial z}T_{22}-\frac{\partial v}{\partial x}T_{11}+T_{12} &= \frac{\partial u}{\partial z},\\
u\frac{\partial T_{22}}{\partial x}+v\frac{\partial T_{22}}{\partial z}-2\frac{\partial v}{\partial x}T_{12}-2\frac{\partial v}{\partial z}T_{22}+T_{22} &= 2\frac{\partial v}{\partial z},\\
\frac{\partial u}{\partial x}+\frac{\partial v}{\partial z} &= 0,\\
\frac{\partial^2 T_{11}}{\partial x\partial z}+\frac{\partial^2 T_{12}}{\partial z^2} &= 0.
\end{aligned} \tag{6.25}$$

A similar rescaling of the equations can be carried out for other constitutive models; in [31], the PTT and Giesekus models were considered. Since the viscometric behavior of these models is very different from the UCM model, different scalings arise: For the PTT model, the boundary layer thickness is of order $W^{-1/3}$, and for the Giesekus model, it is of order $W^{-1/2}$. Compared to the UCM model, these models therefore have boundary layers which sharpen much less rapidly as the Weissenberg number increases, and indeed numerical simulations with these models have generally been much easier to do than with the UCM.

We can reexpress the stress components in a different basis in an analogous fashion to that in the previous section. This puts the boundary layer equations into a somewhat simpler form, which has the added advantage that it also applies on a curved boundary [93]. We set

$$T_{11} = \lambda u^2,\ T_{12} = \lambda uv + \mu,\ T_{22} = \frac{\nu}{u^2} + \frac{2\mu v}{u} + \lambda v^2 - 1. \tag{6.26}$$

With these substitutions, (6.25) assumes the form

$$\begin{aligned} u\lambda_x + v\lambda_z + \lambda - \frac{2\mu u_z}{u^2} &= 0, \\ u\mu_x + v\mu_z + \mu - \frac{\nu u_z}{u^2} &= 0, \\ u\nu_x + v\nu_z + \nu &= u^2, \\ u_x + v_z &= 0, \\ \left[u(\lambda u)_x + v(\lambda u)_z + \mu_z\right]_z &= 0. \end{aligned} \tag{6.27}$$

We can achieve a further simplification by introducing the streamfunction as an independent variable in place of the (rescaled) distance z from the wall. The streamfunction ψ is defined by

$$u = \psi_z,\ v = -\psi_x. \tag{6.28}$$

If u is nonzero, we can solve for z in terms of x and ψ and express all variables in terms of x and ψ. Derivatives transform as follows:

$$\begin{aligned} \frac{\partial f}{\partial x}\bigg|_{z=\text{const.}} &= \frac{\partial f}{\partial x}\bigg|_{\psi=\text{const.}} - v\frac{\partial f}{\partial \psi}, \\ \frac{\partial f}{\partial z} &= u\frac{\partial f}{\partial \psi}. \end{aligned} \tag{6.29}$$

Using these relationships, we transform (6.27) into the system

$$\begin{aligned} u\lambda_x + \lambda - \frac{2\mu u_\psi}{u} &= 0, \\ u\mu_x + \mu - \frac{\nu u_\psi}{u} &= 0, \\ u\nu_x + \nu &= u^2, \\ \left[u(\lambda u)_x + u\mu_\psi\right]_\psi &= 0. \end{aligned} \tag{6.30}$$

In this form, the equations remain valid on a curved boundary if we interpret x as the distance along the wall and u as the absolute value of the velocity.

Of course the equations (6.30) are still a complicated nonlinear system of PDEs, which is by no means easy to solve. There are two situations where we can simplify the equations further and actually obtain solutions: similarity solutions and the linearization at uniform shear flow.

A similarity solution of (6.30) can be found by the ansatz

$$\xi = x^{-\alpha}\psi,\ u = x\tilde{u}(\xi),\ \lambda = x^{2-2\alpha}\tilde{\lambda}(\xi),\ \mu = x^{2-\alpha}\tilde{\mu}(\xi),\ \nu = x^2\tilde{\nu}(\xi). \tag{6.31}$$

We substitute this ansatz into (6.30) and suppress the tildes. This yields the system

$$\begin{aligned}
(2-2\alpha)u\lambda - \alpha\xi u\lambda' + \lambda &= \frac{2\mu u'}{u}, \\
(2-\alpha)u\mu - \alpha\xi u\mu' + \mu &= \frac{\nu u'}{u}, \\
2u\nu - \alpha\xi u\nu' + \nu &= u^2, \\
\left[u((3-2\alpha)\lambda u - \alpha\xi(\lambda' u + u'\lambda) + \mu')\right]' &= 0.
\end{aligned} \tag{6.32}$$

We have thus obtained a system of ordinary differential equations. Still there are no obvious explicit solutions, and solutions have to be found numerically. Nevertheless, numerical integration of a system of ODEs is a far less challenging task than solving a system of PDEs. We shall see in the next chapter how these similarity solutions play a role in the reentrant corner problem.

Next we consider spatially periodic solutions of (6.30) that are small perturbations of plane Couette flow. In our variables, plane Couette flow has the form $u = \psi^{1/2}$, $\nu = \psi$, $\mu = 1/2$, and $\lambda = 1/(2\psi)$. To look at small periodic perturbations of this solution, we set

$$\begin{aligned}
u(x,\psi) &= \psi^{1/2}(1+\tilde{u}(\psi)e^{ix}), \\
\nu(x,\psi) &= \psi + \tilde{\nu}(\psi)e^{ix}, \\
\mu(x,\psi) &= \frac{1}{2} + \tilde{\mu}(\psi)e^{ix}, \\
\lambda(x,\psi) &= \frac{1}{2\psi} + \tilde{\lambda}(\psi)e^{ix}.
\end{aligned} \tag{6.33}$$

We insert this into (6.30) and linearize. The resulting equations can be combined into the single equation

$$\begin{aligned}
&\frac{i\psi\tilde{u}'}{i\psi^{1/2}+1} + \frac{i\psi\tilde{u}'}{(i\psi^{1/2}+1)^2} + \frac{i\tilde{u}}{2} + \frac{i\tilde{u}}{(i\psi^{1/2}+1)^3} \\
&+\psi^{1/2}\frac{d}{d\psi}\left[\frac{\tilde{u}}{(i\psi^{1/2}+1)^2} + \frac{\psi\tilde{u}'}{i\psi^{1/2}+1}\right] = C,
\end{aligned} \tag{6.34}$$

where C is an integration constant independent of ψ.

After substituting $y = \psi^{1/2}$ and some algebraic manipulations, this equation can be simplified to

$$\frac{d}{dy}\left[\frac{y\tilde{u}'}{2(iy+1)}\right] + i\tilde{u} + \tilde{u}' = 2C. \tag{6.35}$$

Here the prime now denotes the derivative with respect to y. We have the obvious solution

$$\tilde{u} = \frac{2C}{i}, \tag{6.36}$$

and it can be shown that this is the only solution which has physically acceptable behavior at the wall $y = 0$ as well as at infinity. For the equations linearized about parallel shear flow, we therefore have a solution with an arbitrary periodic wall-shear rate. We note that the behavior of (6.36) does not change across the boundary layer. In contrast, the stress components change drastically. For instance, if we use (6.36) in the boundary layer equations and compute the stresses, we find

$$\tilde{\nu} = \frac{4C\psi}{i(i\psi^{1/2} + 1)}, \tag{6.37}$$

which is proportional to ψ if ψ is small but proportional to $\psi^{1/2}$ if ψ is large. A similar change occurs in the behavior of the other stress components. The explicit solution for the linearized problem can be used to establish existence of solutions for the nonlinear equations by means of an implicit function argument [96].

We note that in the derivation of the boundary layer equations we assume a nonzero shear rate at the wall. If there are stagnation points at the wall where the shear rate is zero, the analysis breaks down. The behavior of high Weissenberg number flows near such stagnation points and the streamlines emanating from them remains an important open problem. In [95], the flow past a cylinder is analyzed under the presumption that the velocity field near the upstream and downstream stagnation points on the cylinder wall is qualitatively like the Newtonian velocity field. The results of this asymptotic analysis, however, do not appear to agree with the numerical evidence, which suggests that the assumption of a Newtonian-like velocity field is incorrect. This issue warrants further analysis and also more detailed numerical investigation of the local structure of the velocity field near the stagnation points.

Chapter 7

Reentrant Corner Singularities

Many flow problems of interest involve stress singularities at corners of the flow domain. An example is the flow into a contraction as in Figure 1.6. At the point where the flow domain narrows, there is a "reentrant" 270° corner. The resolution of the stress singularity associated with this corner is a major numerical problem. Other examples of corner singularities arise where a free surface attaches to a wall as in the die swell problem or where discontinuous data are prescribed as in the driven-cavity flow.

In this chapter, we shall confine the discussion to corner singularities between two solid walls where the no-slip boundary condition holds. In the Newtonian case, a formal analysis of this problem was first given by Dean and Montagnon [15], and there is by now an extensive mathematical literature concerned with existence and regularity questions for flows with corners (for a review of some of this work, see, e.g., Chapter 7 of [25]). We shall briefly review the Newtonian situation.

7.1 The Newtonian case

The main difference between the Newtonian and non-Newtonian situations is that the corner singularity for the Newtonian fluid is described by a linear problem. Suppose r is the distance from the corner. If the velocity is proportional to r^α, then the viscous stresses behave like $r^{\alpha-1}$, while the Reynolds stresses behave like $r^{2\alpha}$. As long as α is positive, the viscous stresses will always dominate over the Reynolds stresses, and consequently the local flow near the corner is described by the Stokes equation

$$\Delta \mathbf{v} - \nabla p = 0, \quad \operatorname{div} \mathbf{v} = 0. \tag{7.1}$$

In two space dimensions, we can use a streamfunction

$$\mathbf{v} = \left(\frac{\partial \psi}{\partial y}, -\frac{\partial \psi}{\partial x} \right), \tag{7.2}$$

and the Stokes equation becomes equivalent to the biharmonic equation

$$\Delta^2 \psi = 0. \tag{7.3}$$

To study the local behavior near a corner, we look for solutions in a sector of the plane given in polar coordinates by $0 < r < \infty$, $0 < \phi < \alpha$. The biharmonic equation in polar coordinates has the form

$$\left(\frac{\partial^2}{\partial r^2} + \frac{1}{r}\frac{\partial}{\partial r} + \frac{1}{r^2}\frac{\partial^2}{\partial \phi^2} \right)^2 \psi = 0, \tag{7.4}$$

and we need to satisfy the boundary conditions

$$\psi = \frac{\partial \psi}{\partial \phi} = 0 \tag{7.5}$$

at $\phi = 0$ and $\phi = \alpha$. We separate variables and set

$$\psi(r, \phi) = r^\lambda g(\phi). \tag{7.6}$$

The biharmonic equation becomes

$$\left((\lambda - 2)^2 + \frac{d^2}{d\phi^2} \right) \left(\lambda^2 + \frac{d^2}{d\phi^2} \right) g = 0, \tag{7.7}$$

and the boundary conditions are

$$g(0) = g'(0) = g(\alpha) = g'(\alpha) = 0. \tag{7.8}$$

The general solution of the differential equation is

$$g(\phi) = A\cos(\lambda\phi) + B\sin(\lambda\phi) + C\cos((\lambda - 2)\phi) + D\sin((\lambda - 2)\phi), \tag{7.9}$$

as long as λ is not equal to 0, 1, or 2. The requirement that there be a nontrivial solution satisfying the boundary conditions yields an eigenvalue problem for λ. The equation for λ is

$$\lambda(\lambda - 2) - (\lambda - 1)^2 \cos(2\alpha) + \cos(2\alpha(\lambda - 1)) = 0. \tag{7.10}$$

We are interested in solutions that have finite velocity at the corner, so we require that $\lambda > 1$. It then turns out that the smallest eigenvalue determined by (7.10) is less than 2 if $\alpha > \pi$ and has real part greater than 2 if $\alpha < \pi$ (for sufficiently small α, the eigenvalue with smallest real part is complex). Figure 7.1 shows the value of λ for $\pi < \alpha < 2\pi$. For the case of a 270° corner, which occurs in the contraction flow, we have $\lambda = 1.544\ 48$.

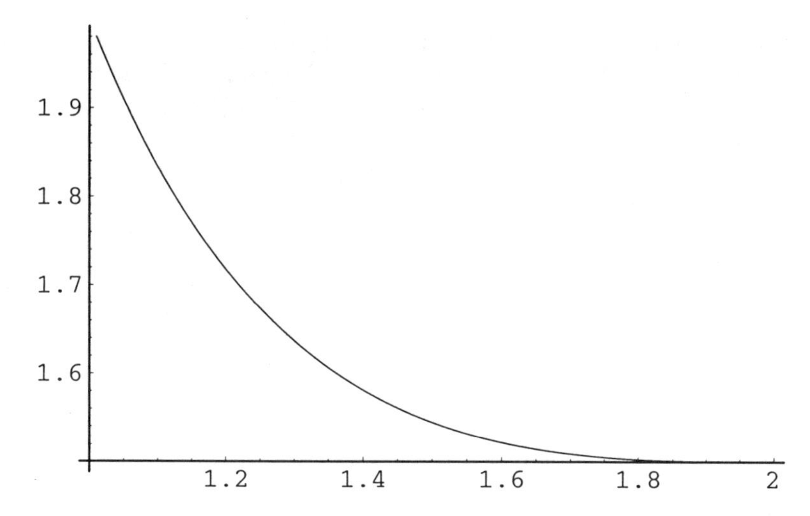

Figure 7.1: Exponent for corner singularity. The plot shows λ as a function of α/π.

The fundamental difference between the two cases is that for $\alpha < \pi$, the velocity gradient, and hence the viscous stresses at the corner, is zero. In the presence of non-Newtonian effects, this makes the local Weissenberg number at the corner equal to zero, and we can expect a corner behavior which is dominated by the Newtonian term. If, on the other hand, $\alpha > \pi$, then the velocity gradient and viscous stresses are infinite at the corner. In that case, the local Weissenberg number is infinite and the non-Newtonian corner behavior is fundamentally different from the Newtonian case.

7.2 The UCM model

For models of viscoelastic flow such as the UCM model, the description of the corner singularity is nonlinear. Moreover, in contrast to power law fluids, for example, the corner behavior cannot be described by a function of the form $r^\lambda g(\phi)$. Indeed, even if we impose a velocity field of this form, the stresses are not of the same form [83]. The reason for this is the appearance of boundary layers on the walls, where the stresses behave differently than in the remainder of the flow domain.

In this section, we shall use the ideas of the previous chapter to derive a formal description of the reentrant corner singularity in terms of matched asymptotics. We assume that the flow in the region away from the walls is described by the Euler equations, while near the wall we have a similarity solution of the boundary layer equations. A crucial assumption in such a description is that there are no streamlines separating from the corner. Actually, there are experiments on viscoelastic contraction flows as well as some numerical simulations (see, e.g., [14]) that suggest that a "lip vortex" attached to the reentrant corner exists in some situations (for an experiment, see Figure 7.2, which is taken from [10]). This feature, however, has not been found for the UCM or Oldroyd B model; it seems to require a different constitutive theory. The effect of separating streamlines on the nature of the corner singularity is an important issue that has not yet been addressed. Of course separating streamlines are also an important feature in other flows with corner singularities, such as the die swell and the driven-cavity problems.

We shall focus on the case of a 270° corner as occurs in contraction flow. Since the stresses at the corner are infinite, we expect high Weissenberg number asymptotics to apply locally, no matter what the nominal Weissenberg number of the overall flow is. Since we are interested in a local solution, we assume the flow domain is given by $0 < r < \infty$, $0 < \theta < 3\pi/2$ in polar coordinates. Due to the self-similar nature of this geometry, we can scale the Weissenberg number out of the problem and we have the equations

$$\begin{aligned} \operatorname{div} \mathbf{T} - \nabla p &= 0, \\ \operatorname{div} \mathbf{v} &= 0, \\ (\mathbf{v}\cdot\nabla)\mathbf{T} - (\nabla\mathbf{v})\mathbf{T} - \mathbf{T}(\nabla\mathbf{v})^T + \mathbf{T} &= \nabla\mathbf{v} + (\nabla\mathbf{v})^T. \end{aligned} \tag{7.11}$$

Away from the walls $\theta = 0$ and $\theta = 3\pi/2$, we expect to be able to neglect the linear terms in the constitutive equation, and we end up with the Euler equations as in the preceding chapter. If there are no separating streamlines emanating from the corner, then all fluid particles originate in a region near the upstream wall where the flow is viscometric. In this viscometric flow, the dominant stress component is in the direction of the streamline. It follows from (6.5) that the dominant stress component remains in the direction of the streamline as the particle rounds the corner. This implies that, in the notation of section 6.1, $\mathbf{u}$ is parallel to $\mathbf{v}$ and $\phi(\rho) = 0$. We may then set $\rho^{1/2}\mathbf{u} = \mathbf{u}^*$, and we end up with the equations

$$\begin{aligned} (\mathbf{u}^*\cdot\nabla)\mathbf{u}^* &= \nabla p, \\ \operatorname{div}\mathbf{u}^* &= 0. \end{aligned} \tag{7.12}$$

These are the steady incompressible Euler equations. If we introduce a stream-function by

$$\mathbf{u}^* = (\psi_y^*, -\psi_x^*), \tag{7.13}$$

then the general solution of the steady Euler equations is given by

$$\Delta\psi^* = q(\psi^*) \tag{7.14}$$

with an arbitrary function q. For the behavior at the corner, we have $\Delta\psi^* = 0$ at leading order, unless the function q has a singularity as $\psi^* \to 0$. Such a singularity, however, would lead to singular behavior of the solution at the walls as well as the corner, and it seems doubtful that such solutions are physically relevant. Hence we are looking for a potential flow solution. Since $\mathbf{u}^*$ is parallel

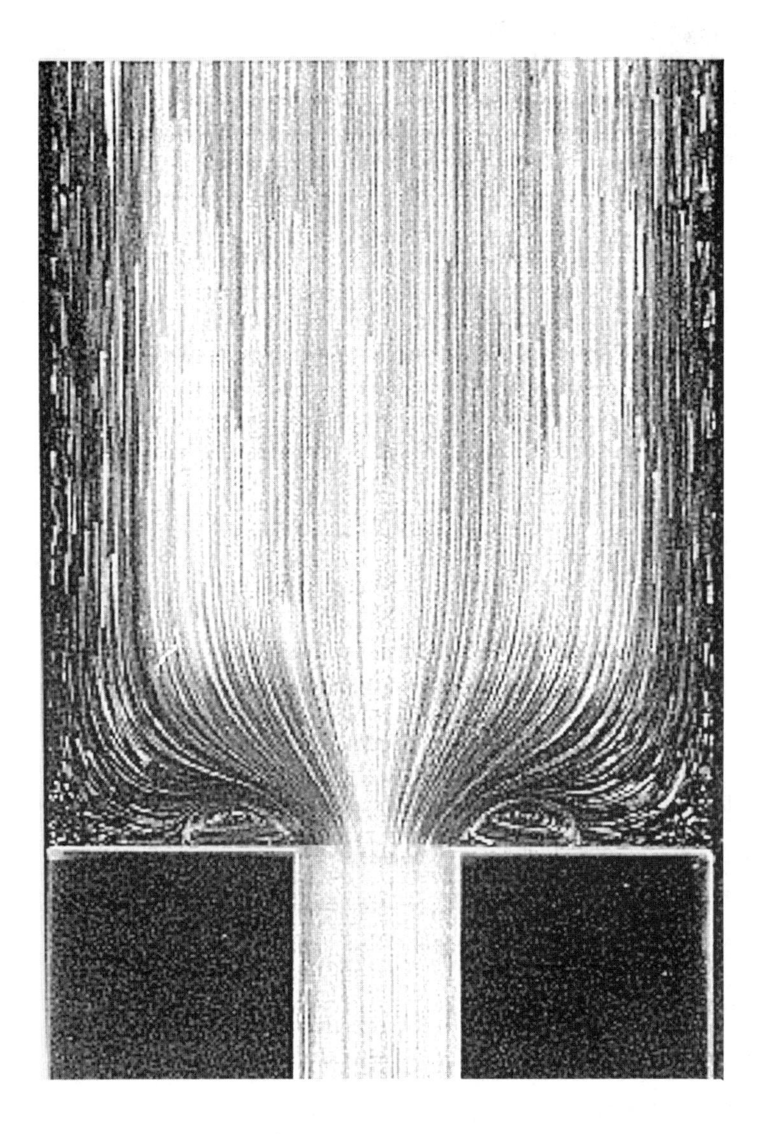

Figure 7.2: Lip vortex in contraction flow (Figure 3.7c of [10]). Reprinted from D.V. Boger and K. Walters, *Rheological Phenomena in Focus*, 1993, 42, with permission from Elsevier Science.

to $\mathbf{v}$, we have the boundary condition $\psi^* = 0$ just as in the case of inviscid flow. The potential flow solution for the 270° corner is given by

$$\psi^* = r^{2/3} \sin\left(\frac{2}{3}\theta\right). \tag{7.15}$$

Of course ψ^* is the streamfunction for $\mathbf{u}^*$, which is not the true velocity field. However, $\mathbf{u}^*$ and $\mathbf{v}$ are parallel, so they have the same streamlines. Consequently, the true streamfunction must be a function of ψ^*: $\psi = h(\psi^*)$. For the local behavior at the corner, only the behavior of $h(\psi^*)$ as $\psi^* \to 0$ is relevant. Let us assume a power behavior $h(\psi^*) = (\psi^*)^n$. Then we have

$$\psi = r^{2n/3} \sin^n \left(\frac{2}{3}\theta\right). \tag{7.16}$$

This description of the flow breaks down in the viscometric region near the walls. The transition from Euler to viscometric behavior occurs when the "stretch rate"

$$\frac{\partial v_r}{\partial r} = \frac{\partial}{\partial r}\left(\frac{1}{r}\psi_\theta\right) \tag{7.17}$$

is of order 1 [34]. This is the case when

$$r^{(2n-6)/3}\theta^{n-1} \tag{7.18}$$

is of order 1. At this point, the shear rate $\psi_{\theta\theta}/r^2$ is of order

$$r^{(2n-6)/3}\theta^{n-2}, \tag{7.19}$$

and viscometric stresses at this shear rate would be of the order of magnitude

$$r^{(4n-12)/3}\theta^{2n-4}. \tag{7.20}$$

On the other hand, the stresses found from the Euler solution above are given by $\mathbf{u}^*(\mathbf{u}^*)^T$, i.e., the dominant stress component is of order $|\nabla\psi^*|^2$, i.e., of order $r^{-2/3}$. If there is to be a transition from viscometric to Euler behavior, then these orders of magnitudes ought to match each other, that is, when

$$r^{(2n-6)/3}\theta^{n-1} \tag{7.21}$$

is of order 1, then we should have

$$r^{(4n-12)/3}\theta^{2n-4} \sim r^{-2/3}. \tag{7.22}$$

This condition leads to $n = 7/3$.

Near each wall, the solution has a stress boundary layer. From the above considerations, we know that the boundary layer must occupy the region where

$$r^{-4/9}\theta^{4/3} \tag{7.23}$$

is of order 1 (set $n = 7/3$ in (7.21)). Hence θ is of order $r^{1/3}$, and consequently ψ is of order $r^{7/3}$. Consequently, it is natural to look for a similarity solution of the boundary layer equations, i.e., a solution of (6.32) with the choice of $\alpha = 7/3$. At the wall, we need to have viscometric behavior, i.e.,

$$u \sim \xi^{1/2},\ \lambda \sim \frac{1}{2\xi},\ \mu \sim \frac{1}{2},\ \nu \sim \xi. \tag{7.24}$$

For $\xi \to \infty$, the solution needs to match with the potential flow. That is, the asymptotic behavior of the boundary layer solution as $\xi \to \infty$ needs to agree with the asymptotic behavior of the potential flow solution as we approach the wall. This condition translates into the requirement that

$$u \sim C_1\xi^{4/7},\ \lambda \sim C_2\xi^{-8/7},\ \mu = o(1),\ \nu = o(\xi^{8/7}) \tag{7.25}$$

as $\xi \to \infty$. At this point, there is no analysis of (6.32) to establish the existence of such solutions, and we have to rely on numerical evidence. For the boundary layer upstream from the corner, numerical solutions were found in [89]. There is a crucial difference between the upstream and downstream situations. For the upstream case, we can impose the viscometric stresses at the wall and then integrate (6.32) as an initial value problem. For the downstream situation, however, this procedure leads to an ill-posed problem. Hence we need to consider (6.32) as a two-point boundary value problem with conditions imposed both at zero and infinity. This problem has not yet been analyzed.

7.3 Instability of stress integration

Every method of numerical solution of viscoelastic flows must in some form involve the integration of the constitutive relation along streamlines. As we shall see, this integration is inherently unstable in the downstream region of a corner singularity. The source of this instability is a feature of the equations themselves and has nothing to do with the choice of numerical method. We can see what happens by investigating the integration of (6.5), which we restate for convenience:

$$\begin{aligned}
(\mathbf{v}\cdot\nabla)\lambda + 2\mu\,\mathrm{div}\,\mathbf{w} &= 0,\\
(\mathbf{v}\cdot\nabla)\mu + \nu\,\mathrm{div}\,\mathbf{w} &= 0,\\
(\mathbf{v}\cdot\nabla)\nu &= 0.
\end{aligned} \tag{7.26}$$

These equations represent the high Weissenberg number limit of the constitutive equation, where the stress has been written in terms of components aligned with the velocity field. If we look now at the last equation, it simply says that ν is constant along streamlines. This does not sound like there is instability, but note that the actual stress associated with ν is $\nu\mathbf{w}\mathbf{w}^T$, and $\mathbf{w}$ becomes infinite as the velocity approaches zero. This is the case along streamlines approaching the wall downstream from a corner. Worse yet, $\mathrm{div}\,\mathbf{w}$ becomes infinite even more rapidly than $\mathbf{w}$ itself, and so the terms involving $\mathrm{div}\,\mathbf{w}$ will also lead to a large amplification of errors.

As we saw in the preceding section, the stresses near the corner are dominated by the λ component, and the μ and ν components are much smaller. The instability leads to a stress growth downstream, but it is the μ and ν components that are being magnified, and they start from a small value. The picture changes, however, if numerical errors enter into consideration. A typical numerical discretization of the constitutive law will not have the triangular, partially decoupled form of (6.5), and as a consequence numerical errors from the dominant λ

component will pollute the μ and ν components. The closer we get to the corner, the more dominant the λ component becomes, and eventually the numerical errors in the μ and ν components become larger than the true values. The instability then magnifies these numerical errors downstream. A remedy for this is to choose a numerical method that respects the decoupled character of (6.5). The dramatic difference that this can make was demonstrated in [84]. The problem under consideration there is simply the integration of stresses in a given velocity field, which is taken to be the Newtonian velocity field for a 270° corner. The stress integration along each streamline then involved solving a system of ODEs. Each streamline was parametrized by the angle θ in polar coordinates, and then the ODEs were discretized using the implicit Euler scheme and equidistant steps. Viscometric stresses were imposed as an initial condition at $\theta = 10^{-4}$, and 1000 Euler steps were used to discretize the interval $[10^{-4}, 3\pi/2]$. The stresses were expressed in two different ways: First, using conventional polar components T_{rr}, $T_{r\theta}$, and $T_{\theta\theta}$, the constitutive equation was transformed into a system for these components and then discretized. For the second calculation, the stresses were expressed in terms of λ, μ, and ν as defined before; the constitutive equation was expressed in those components and then discretized. Only after the numerical solution were the stresses converted back to polar components so we could compare the results of the two calculations. The tables below show the computed stresses at various angles for a given small value of the streamfunction ($\psi = 10^{-5}$).

θ	T_{rr}	$T_{r\theta}$	$T_{\theta\theta}$
0.5043	3 948.17	−1 560.17	617.77
1.0038	4 623.16	−3 989.10	3 451.65
2.0028	606.10	−2 652.46	12 042.40
3.0018	1 950.25	4 316.09	9 841.00
4.0008	4 913.58	2 516.87	1 439.85
4.5003	7 598.64	−105.27	97.82
4.6841	234 034.67	−2 825.00	34.85

θ	T_{rr}	$T_{r\theta}$	$T_{\theta\theta}$
0.5043	3 464.48	−1 355.36	529.24
1.0038	4 031.09	−3 455.29	2 960.14
2.0028	525.10	−2 338.47	10 349.47
3.0018	1 616.61	3 718.82	8 548.84
4.0008	4 024.78	2 234.32	1 239.24
4.5003	2 069.83	312.50	46.54
4.6841	962.44	−8.04	−0.37

The first table clearly shows the effect of the downstream instability and large spurious stresses, which are completely avoided in the second table.

A similar calculation with the PTT model [94] did not show a downstream growth of errors. The reason for this is that the boundary layers in the PTT model have a different scaling and are much wider than in the UCM model. As a result, particles enter the boundary layer and stress relaxation takes over before the downstream instability has had a chance to fully manifest itself.

Chapter 8

Instabilities

Flow instabilities, and the patterns emerging from them, are a major topic of interest in Newtonian flows [7]. Indeed, classical problems of hydrodynamic instability such as the Bénard and Taylor problem have inspired much of the modern work in dynamical systems. Viscoelastic flows introduce new physical effects which can become a source of flow instabilities. Examples of viscoelastic flows that can exhibit instabilities include shear flows with curved streamlines (e.g., flow between rotating cylinders or cone-and-plate flow), elongational flows (e.g., fiber spinning), contraction flows, and flows with fluid interfaces (e.g., co-extrusion). We shall not attempt to review the extensive literature in this rapidly growing field; Larson's review article [49] is a good place to start to get an overview of this area.

The focus of this chapter will be on the issues that instabilities present from the point of view of mathematical analysis. We shall begin with a review of the traditional dynamical systems approach to the study of stability and bifurcation, i.e., the methods to determine when instabilities occur and how to simplify the equations near points where they do. These techniques, originally established for systems of ordinary differential equations, have been extended to many partial differential equations, including those for Newtonian fluid mechanics. The hyperbolic nature of the constitutive laws for viscoelastic fluids, however, presents new and serious issues. We shall review some recent work aimed at addressing those issues.

8.1 Analysis of instabilities

Consider a system of ordinary differential equations

$$\dot{u} = h(u, \lambda), \tag{8.1}$$

where $u \in \mathbb{R}^n$, λ is a real parameter, and h is a smooth function of both arguments. For physical applications, we should think of u as representing the solution to a flow problem and λ as a parameter on which the problem depends, e.g., the Weissenberg number. We shall assume that (8.1) has a stationary solution $u_0(\lambda)$: $h(u_0(\lambda),\lambda) = 0$, and we look for small perturbations of this solution: $u = u_0(\lambda) + v$. We can then rewrite (8.1) as

$$\begin{aligned} \dot{v} &= h(u_0(\lambda) + v, \lambda) = h(u_0(\lambda) + v, \lambda) - h(u_0(\lambda), \lambda) \\ &= C(\lambda)v + Q(v, \lambda). \end{aligned} \tag{8.2}$$

Here $C(\lambda) := Dh(u_0(\lambda), \lambda)$ denotes the Jacobian matrix $Dh_{ij} = \partial h_i/\partial x_j$, and the remainder term Q has the property that $\|Q(v,\lambda)\| \leq C\|v\|^2$ for $\|v\|$ sufficiently small.

We say that the stationary solution $u_0(\lambda)$ is stable to small disturbances if $\|v(t)\| \to 0$ as $t \to \infty$ provided that $\|v(0)\|$ is sufficiently small. It is reasonable to expect that stability to small disturbances is determined by the linearized system

$$\dot{v} = C(\lambda)v. \tag{8.3}$$

As is well known, the zero solution of this system is stable if all eigenvalues of the matrix $C(\lambda)$ are in the left half-plane and unstable if at least one eigenvalue is in the right half-plane. Moreover, it can be shown that the stability for the linearized system does indeed determine stability to small disturbances for the full nonlinear system.

8.2 Center manifolds

At the onset of instability, we have one or several eigenvalues of $C(\lambda)$ crossing the imaginary axis as λ passes through a critical value λ_0. The number of eigenvalues and eigenvectors that are associated with this change is typically small. Hence, bifurcation problems usually involve systems where the linearization has a very large, and possibly infinite-dimensional, stable part and a small number of "critical" that change from stable to unstable as the bifurcation parameter exceeds a threshold. The central idea of bifurcation theory is that the dynamics of the system near the onset of instability is governed by the evolution of these critical modes, while the stable modes follow in a passive fashion—they are "enslaved." The center manifold theorem is the rigorous formulation of this idea; it allows us to reduce a large problem to a small and manageable one.

We shall state the theorem for systems of ODEs; of course the real physical applications are PDEs. We state our system in a form where the stable and critical modes have been separated out. Specifically, we let $x \in \mathbb{R}^m$ and $y \in \mathbb{R}^n$, and we consider the system

$$\begin{aligned} \dot{x} &= A(\lambda)x + f(x, y, \lambda), \\ \dot{y} &= B(\lambda)y + g(x, y, \lambda). \end{aligned} \tag{8.4}$$

We make the following assumptions:

1. f and g are smooth (C^∞) functions, and they are quadratic at the origin, i.e.,

$$\|f(x, y, \lambda)\| + \|g(x, y, \lambda)\| \leq C(\|x\|^2 + \|y\|^2) \tag{8.5}$$

for small $\|x\|$ and $\|y\|$.

2. A and B depend smoothly on λ. All eigenvalues of $A(0)$ are purely imaginary, while all eigenvalues of $B(0)$ have negative real parts.

Theorem 3 *If assumptions 1 and 2 hold, then, in some neighborhood of* $x = 0$, $\lambda = 0$, *there exists a* C^∞*-smooth manifold of the form* $y = \phi(x, \lambda)$, *called the center manifold, with the following properties:*

A. $\phi(0, \lambda) = 0$ *and* $\partial\phi/\partial x_i(0, 0) = 0$ *for* $i = 1, ..., m$.

B. Every solution of (8.4) that starts on the center manifold remains on it, i.e., if $y(0) = \phi(x(0), \lambda)$, *then* $y(t) = \phi(x(t), \lambda)$ *for all* t.

C. Every solution of (8.4) that remains small for positive time approaches the center manifold. That is, there exists an $\epsilon > 0$ *such that if* $|\lambda| < \epsilon$ *and* $\|x(t)\| + \|y(t)\| < \epsilon$ *for all* $t > 0$, *then* $\|y(t) - \phi(x(t), \lambda)\| \to 0$ *as* $t \to \infty$. *Every solution that satisfies* $\|x(t)\| + \|y(t)\| < \epsilon$ *for all* t *(positive and negative) is on the center manifold. Moreover, the stability of such a solution is determined by its stability within the center manifold.*

Property C implies that if we are looking for steady or periodic solutions bifurcating from the trivial solution $x = 0$, $y = 0$, then we can confine our search to the center manifold. This is a substantial simplification, since in applications m is typically very small (1 or 2 in the situations considered below), while n is large or even infinite.

8.3 Simple eigenvalue bifurcation

We shall now consider how the center manifold theorem can be used to study the simplest bifurcations. First, we consider the case where a simple eigenvalue of the linearized problem crosses zero. That is, we consider a system of the form

$$\dot{u} = A(\lambda)u + f(u, \lambda), \tag{8.6}$$

where u is a vector, the matrix $A(0)$ has a simple eigenvalue 0 and the remainder of the eigenvalues are in the left half-plane, and f is of quadratic or higher order: $\|f(u, \lambda)\| \leq C\|u\|^2$.

For λ close to zero, there exists an eigenvalue $\sigma(\lambda)$ which is close to zero; we assume that, as λ goes through zero, this eigenvalue crosses from negative to positive with nonvanishing speed:

$$\frac{d}{d\lambda}\sigma(\lambda)|_{\lambda=0} > 0. \tag{8.7}$$

Let $a(\lambda)$ be the eigenvector of $A(\lambda)$ with eigenvalue $\sigma(\lambda)$, and let $b(\lambda)$ be the adjoint eigenvector, i.e.,

$$A(\lambda)a(\lambda) = \sigma(\lambda)a(\lambda), \quad (b(\lambda), (A(\lambda) - \sigma(\lambda))u) = 0. \tag{8.8}$$

We normalize $a(\lambda)$ and $b(\lambda)$ in such a way that $(b(\lambda), a(\lambda)) = 1$. It can then be shown that

$$\frac{d}{d\lambda}\sigma(\lambda)|_{\lambda=0} = \left(b(0), \frac{dA}{d\lambda}(0)a(0)\right). \tag{8.9}$$

We now decompose u into a component proportional to the eigenvector $a(\lambda)$ and a component perpendicular to $b(\lambda)$:

$$u = xa(\lambda) + y, \quad (b(\lambda), y) = 0. \tag{8.10}$$

Here x is a scalar variable. The system (8.6) now takes on the form

$$\begin{aligned} \dot{x} &= \sigma(\lambda)x + g(x, y, \lambda), \\ \dot{y} &= B(\lambda)y + h(x, y, \lambda). \end{aligned} \tag{8.11}$$

Here $B(\lambda)$ is simply the restriction of $A(\lambda)$ to the subspace characterized by the condition $(b(\lambda), y) = 0$; by assumption all eigenvalues of $B(\lambda)$ are in the left half-plane if λ is small. The nonlinear terms g and h are the projections of the original nonlinearity f:

$$\begin{aligned} g(x, y, \lambda) &= (b(\lambda), f(xa(\lambda) + y, \lambda)), \\ h(x, y, \lambda) &= f(xa(\lambda) + y, \lambda) - g(x, y, \lambda)a(\lambda). \end{aligned} \tag{8.12}$$

According to the center manifold theorem, there is a center manifold of the form $y = \phi(x, \lambda)$. For the following, we only need the quadratic approximation to the center manifold. We expand g and h as follows:

$$\begin{aligned} g(x, y, \lambda) &= \alpha_1(\lambda)x^2 + \alpha_2(\lambda)x^3 + (p_1(\lambda), y)x \\ &\quad + O(|x|^4 + \|y\|^2 + |x|^2\|y\|), \\ h(x, y, \lambda) &= p_2(\lambda)x^2 + O(|x|^3 + |x|\|y\| + \|y\|^2). \end{aligned} \tag{8.13}$$

Here the α_i are scalars and the p_i are vectors. The quadratic approximation to the center manifold is

$$\phi(x, \lambda) = -(B(\lambda) - 2\sigma(\lambda))^{-1}p_2(\lambda)x^2 + O(x^3), \tag{8.14}$$

and by inserting this back into the differential equation for x, we find the following differential equation on the center manifold, including terms up to order x^3:

$$\dot{x} = \sigma(\lambda)x + \alpha_1(\lambda)x^2 + (\alpha_2(\lambda) - (p_1(\lambda), (B(\lambda) - 2\sigma)^{-1}p_2(\lambda)))x^3 + O(|x|^4). \tag{8.15}$$

We write (8.15) in the form

$$\dot{x} = \sigma(\lambda)x + \alpha(\lambda)x^2 + \beta(\lambda)x^3 + O(x^4). \tag{8.16}$$

To look for stationary solutions, we set $\dot{x} = 0$. For nonzero solutions, we divide the right-hand side by x, and find

$$\sigma(\lambda) + \alpha(\lambda)x + \beta(\lambda)x^2 + O(x^3) = 0. \tag{8.17}$$

If $\alpha(0) \neq 0$, then the implicit function theorem allows us to solve this equation uniquely for x in a neighborhood of $x = 0$, $\lambda = 0$. The solution is of the form

$$x = x_0(\lambda) = -\frac{\sigma'(0)}{\alpha(0)}\lambda + O(\lambda^2). \tag{8.18}$$

At leading order, x is therefore proportional to λ, that is, a bifurcating solution exists both for $\lambda > 0$ and $\lambda < 0$. The bifurcation is therefore referred to as transcritical. We can next look at the stability of the bifurcating solution. If (8.16) is linearized at $x = x_0$, then the eigenvalue of the linearized problem is

$$\sigma(\lambda) + 2\alpha(\lambda)x_0(\lambda) + 3\beta(\lambda)x_0(\lambda)^2 + O(x_0(\lambda)^3) = -\sigma'(0)\lambda + O(\lambda^2). \tag{8.19}$$

Since $\sigma'(0)$ was assumed positive, the bifurcating solution is stable for $\lambda > 0$ (i.e., precisely where the zero solution is unstable) and unstable for $\lambda < 0$.

Many physical systems have a symmetry which forces the right-hand side of (8.16) to be an odd function of x. Instead of (8.16), we therefore have

$$\dot{x} = \sigma(\lambda)x + \beta(\lambda)x^3 + O(x^5). \tag{8.20}$$

If $\beta(0) \neq 0$, then we can solve for x^2 as a function of λ, that is, stationary solutions are given by

$$x^2 = x_0(\lambda)^2 = -\frac{\sigma'(0)}{\beta(0)}\lambda + O(\lambda^2). \tag{8.21}$$

Bifurcating solutions exist only if $\sigma'(0)\lambda/\beta(0) < 0$, i.e., they exist for $\lambda > 0$ (supercritical bifurcation) if $\beta(0) < 0$ and for $\lambda < 0$ (subcritical bifurcation) if $\beta(0) > 0$. The eigenvalue governing stability of the bifurcating solution turns out to be

$$\sigma(\lambda) + 3\beta(\lambda)x_0(\lambda)^2 + O(x_0(\lambda)^4) = -2\sigma'(0)\lambda + O(\lambda^2). \tag{8.22}$$

This means that supercritical branches are stable while subcritical branches are unstable.

8.4 Hopf bifurcation

Another possibility for an instability is when a pair of complex conjugate eigenvalues crosses the imaginary axis. Again, we shall consider a system of the form

$$\dot{u} = A(\lambda)u + f(u, \lambda). \tag{8.23}$$

We now assume that $A(0)$ has a pair of simple purely imaginary eigenvalues $\pm i\omega_0$, while the rest of the spectrum is in the left half-plane.

In a fashion analogous to before, we denote by $\sigma(\lambda)$ the eigenvalue which passes through $i\omega_0$ as λ passes through 0, and we denote by $a(\lambda)$ the eigenvector and by $b(\lambda)$ the adjoint eigenvector:

$$A(\lambda)a(\lambda) = \sigma(\lambda)a(\lambda), \quad (b(\lambda), (A(\lambda) - \sigma(\lambda))u) = 0. \tag{8.24}$$

We normalize the eigenvectors such that $(b(\lambda), a(\lambda)) = 1$. The analogue of assumption (8.7) is

$$\frac{d}{d\lambda}(\mathrm{Re}\,\sigma(\lambda)) = \mathrm{Re}(b(0), A'(0)a(0)) > 0. \tag{8.25}$$

We introduce the projection operator $P(\lambda)$ as follows:

$$P(\lambda)u = u - (b(\lambda), u)a(\lambda) - (\overline{b(\lambda)}, u)\overline{a(\lambda)}. \tag{8.26}$$

The analogue of the decomposition (8.10) is

$$u = za(\lambda) + \bar{z}\overline{a(\lambda)} + y, \tag{8.27}$$

where $y = P(\lambda)u$ and $(b(\lambda), u) = z$. The scalar z is complex. The analogue of (8.11) is

$$\begin{aligned} \dot{z} &= \sigma(\lambda)z + g(z, y, \lambda), \\ \dot{y} &= B(\lambda)y + h(z, y, \lambda). \end{aligned} \tag{8.28}$$

Here we have

$$\begin{aligned} g(z, y, \lambda) &= (b(\lambda), f(za(\lambda) + \bar{z}\overline{a(\lambda)} + y, \lambda)), \\ h(z, y, \lambda) &= P(\lambda)f(za(\lambda) + \bar{z}\overline{a(\lambda)} + y, \lambda), \\ B(\lambda) &= P(\lambda)A(\lambda). \end{aligned} \tag{8.29}$$

As before, we use the center manifold reduction. We need the center manifold up to quadratic order. The result is

$$y = \phi(z, \lambda) = \psi(\lambda)z^2 + \overline{\psi(\lambda)}\bar{z}^2 + 2\chi(\lambda)z\bar{z} + O(|z|^3). \tag{8.30}$$

We can use (8.30) in the first equation of (8.28) and obtain the following equation for z, up to cubic order:

$$\begin{aligned} \dot{z} &= \sigma(\lambda)z + \alpha_1(\lambda)z^2 + \alpha_2(\lambda)z\bar{z} + \alpha_3(\lambda)\bar{z}^2 \\ &+ \beta_1(\lambda)z^3 + \beta_2(\lambda)|z|^2 z + \beta_3(\lambda)|z|^2\bar{z} + \beta_4(\lambda)\bar{z}^3 + O(|z|^4). \end{aligned} \tag{8.31}$$

By employing a transformation of the form

$$w = z + \gamma_1(\lambda)z^2 + \gamma_2(\lambda)z\bar{z} + \gamma_3(\lambda)\bar{z}^2 + O(|z|^3), \tag{8.32}$$

(8.31) can be reduced to the "Birkhoff normal form" [23]

$$\dot{w} = \sigma(\lambda)w + \beta(\lambda)|w|^2 w + O(|w|^5). \tag{8.33}$$

If we truncate the equation at the cubic order, then we can easily find an explicit periodic solution

$$w(t) = R\exp(i\omega t), \tag{8.34}$$

where

$$\begin{aligned} \mathrm{Re}(\sigma(\lambda)) + R^2\,\mathrm{Re}(\beta(\lambda)) &= 0, \\ \mathrm{Im}(\sigma(\lambda)) + R^2\,\mathrm{Im}(\beta(\lambda)) &= \omega. \end{aligned} \tag{8.35}$$

If higher-order corrections are taken into account, then this explicit solution arises as the first term in an asymptotic approximation, i.e., there is a bifurcating family of periodic solutions parametrized by their amplitude R such that

$$\begin{aligned} \lambda &= -R^2 \frac{\mathrm{Re}(\beta(0))}{\mathrm{Re}(\sigma'(0))} + O(R^4), \\ w(t) &= R\exp(i\omega t) + O(R^3), \\ \omega &= \omega_0 + R^2\left(\mathrm{Im}(\beta(0)) - \frac{\mathrm{Im}(\sigma'(0))\mathrm{Re}(\beta(0))}{\mathrm{Re}(\sigma'(0))}\right) + O(R^4). \end{aligned} \tag{8.36}$$

Since we assumed that $\mathrm{Re}(\sigma'(0)) > 0$, we have a supercritical bifurcation if $\mathrm{Re}(\beta(0)) < 0$ and a subcritical bifurcation if $\mathrm{Re}(\beta(0)) > 0$.

To analyze the stability of the bifurcating periodic solution, it suffices to study the cubic truncation. In the equation

$$\dot{w} = \sigma(\lambda)w + \beta(\lambda)|w|^2 w, \tag{8.37}$$

we set $w = (R+v)\exp(i\omega t)$, where R and ω are as above, and we linearize with respect to v. The resulting linearized equation is

$$\dot{v} + i\omega v = \sigma(\lambda)v + 2\beta(\lambda)R^2 v + \beta(\lambda)R^2\bar{v}. \tag{8.38}$$

Keeping in mind that $\sigma + R^2\beta = i\omega$, we can simplify this to

$$\dot{v} = \beta(\lambda)R^2(v + \bar{v}). \tag{8.39}$$

This system has eigenvalues $2\mathrm{Re}(\beta(\lambda))R^2$ and 0. The zero eigenvalue results from the neutral direction, which allows for a phase shift in the periodic solution. The stability is determined by the other eigenvalue. We find that the branch of periodic solutions is stable if it is supercritical and unstable if it is subcritical.

8.5 Linear stability for PDEs

There are a number of crucial steps in the analysis of stability and bifurcation as outlined in the preceding sections:

1. The reduction of the problem of stability for small disturbances to the study of the linearized equation.

2. The reduction of linear stability to eigenvalues.

3. The reduction of the dynamics near the onset of instability to a low-dimensional center manifold.

All these steps are well established for systems of ODEs and the method has been generalized to a broad class of "parabolic" PDEs, including the equations of Newtonian fluid mechanics. Unfortunately, however, these results do not cover the equations of viscoelastic flows. While we can formally apply the techniques outlined above, a rigorous foundation exists only in fragments at this time. In this section, we shall focus on the linear stability question; I refer to [82] for a version of the center manifold theorem that is applicable to the viscoelastic Bénard problem.

The abstract framework for the study of linear stability is that of strongly continuous semigroups. We consider an initial value problem of the form

$$\dot{u} = Au, \ \ u(0) = u_0, \tag{8.40}$$

where u takes values in a Banach space X and A is a densely defined, closed operator. The Hille–Yosida theorem (see, e.g., [62]) characterizes a class of operators for which the initial value problem is well posed; namely, if there is a $\mu \in \mathbb{R}$ such that the resolvent $(A-\lambda)^{-1}$ exists for $\operatorname{Re}\lambda > \mu$ and satisfies a bound of the form

$$\|(A-\lambda)^{-n}\| \leq \frac{M}{(\operatorname{Re}\lambda - \mu)^n}, \tag{8.41}$$

then there exists a family of bounded linear operators $\exp(At)$, $t \geq 0$, with the following properties:

1. For every $u \in X$, $\exp(At)u$ is continuous in t.

2. $\exp(At)\exp(As) = \exp(A(t+s))$.

3. $\exp(A0) = I$.

4. If $u \in D(A)$, then $\frac{d}{dt}\exp(At)u = Au$.

5. For every $u \in X$, we have

$$\exp(At)u = \lim_{n\to\infty}\left(I - \frac{At}{n}\right)^{-n} u. \tag{8.42}$$

6. The operators $\exp(At)$ satisfy $\|\exp(At)\| \leq M\exp(\mu t)$ with M and μ as in (8.41).

For every strongly continuous semigroup of operators, we can define the growth abscissa or type of the semigroup:

$$\omega(A) = \lim_{t\to\infty}\frac{\ln\|\exp(At)\|}{t}. \tag{8.43}$$

The question of linear stability or instability is then whether $\omega(A)$ is negative or positive. In the case of systems of ODEs, we decided this question by looking at the eigenvalues of A. In infinite dimensions, the spectrum of A need not consist entirely of eigenvalues, but we can allow for this and more generally define the spectral bound

$$r(A) = \sup\{\mathrm{Re}\,\lambda \mid \lambda \in \sigma(A)\}, \tag{8.44}$$

where $\sigma(A)$ denotes the spectrum of A. The question is then whether $\omega(A)$ and $r(A)$ are the same. It is always true that $\omega(A) \geq r(A)$, but it is possible for the inequality to be strict.

If the underlying space X is a Hilbert space, there is the following characterization of $\omega(A)$ [20, 33, 35, 65]:

$$\omega(A) = \inf\{\mu \mid \|(A-\lambda)^{-1}\| \text{ is uniformly bounded for } \mathrm{Re}\,\lambda \geq \mu\}. \tag{8.45}$$

A strategy to prove that $\omega(A) = r(A)$ is therefore to somehow deduce the boundedness of the resolvent in a half-plane from the absence of spectral values. One situation where this can easily be done is if

$$\|(A-\lambda)^{-1}\| \to 0 \tag{8.46}$$

as $\mathrm{Im}\,\lambda \to \infty$. Indeed, this is the case in many applications, e.g., in parabolic PDEs. For hyperbolic PDEs, however, (8.46) does not hold, and indeed it is possible to give examples where $\omega(A) > r(A)$. The following example was given in [85]. Consider the second-order hyperbolic equation

$$u_{tt} = u_{xx} + u_{yy} + e^{ix}u_y, \tag{8.47}$$

with 2π-periodic boundary conditions in each direction. The problem can be put into the abstract form $\dot{z} = Az$, where the underlying function space is $H^1 \times L^2$, and with $z = (u, v)$ we have

$$A(u, v) = (v, u_{xx} + u_{yy} + e^{ix}u_y) \tag{8.48}$$

with domain $D(A) = H^2 \times H^1$. It is shown in [85] that $r(A) = 0$ but $\omega(A) = 1/2$.

For hyperbolic PDEs, there is therefore no hope of establishing a general result to the effect that $\omega(A) = r(A)$, i.e., it is possible to have an onset of instability that is not associated with the spectrum of A. Whether or not this ever happens in viscoelastic flows is not known. For rigorous stability analysis, one can then pursue two approaches:

1. Forget about trying to show $\omega(A) = r(A)$ and ask what additional information, apart from $r(A)$, allows us to determine $\omega(A)$. A result of this type for scalar second-order hyperbolic PDEs was found by Koch and Tataru [45].

2. Look for more restricted problems where $\omega(A) = r(A)$.

In the context of viscoelastic flows, two quite different results of the latter type have been established. The result in [85] concerns parallel shear flows of a fluid with a differential constitutive law of Jeffreys type (no results for fluids of Maxwell type are known at this point). The crucial issue is to obtain a bound on the resolvent of A as $\operatorname{Im}\lambda \to \infty$. In parallel shear flow, due to the translational symmetry, it is possible to separate variables and decompose the operator A into operators A_α associated with different wave numbers α. It turns out that getting an estimate on the resolvent of A_α for fixed α is no problem, and the only issue is for the limit $\alpha \to \infty$. The approach in [85] is based on identifying a "limit problem" for $\alpha \to \infty$. Based on the analysis of this limit problem, it is then possible to find either a bound on the resolvent or prove the existence of eigenvalues.

The result in [90] is of a different nature. Here no symmetry that allows separation of variables is assumed and the flow geometry is quite general. However, the Weissenberg number needs to be assumed sufficiently small. The idea is then to decompose the problem into a "good" part and a "bad" part. The good part has a resolvent that tends to zero as $\operatorname{Im}\lambda \to \infty$ and the resolvent for the bad part can be controlled by an energy estimate as long as the Weissenberg number is small enough. Besides showing that linear stability is determined by the spectrum, [90] also establishes nonlinear stability to small disturbances.

Chapter 9

Change of Type

A change in the type of the governing equations is familiar from gas dynamics. One example of change of type occurs in a van der Waals gas where the pressure is a nonmonotone function of density. In the region of parameter space where the pressure decreases with density, the equations of motion are ill posed, and consequently this becomes a "forbidden region": admissible solutions are to take values only in those regions where the pressure increases with density. The result is a separation into phases. Another kind of change of type occurs in transonic flow. Here the equations of time evolution do not become ill posed, but the equations for steady flow change type, leading to the possibility of discontinuities (shocks) and a change in the nature of boundary conditions which need to be imposed.

In viscoelastic flows, analogues of these phenomena can occur. We begin this chapter with a review of the basic mathematical notions and definitions; this section essentially follows [97]. We then discuss how these notions apply to model equations for viscoelastic flow. Finally we discuss possible connections with phenomena such as melt fracture and delayed die swell.

9.1 Mathematical definitions

An important ingredient of a systematic theory of PDEs is a classification scheme that identifies classes of equations with common properties. The "type" of an equation determines the nature of boundary and initial conditions that may be imposed, the nature of singularities that solutions may have, and the nature of methods that can be used to approximate a solution.

The notation of multi-indices is very convenient in avoiding excessively cumbersome notations in PDEs. A multi-index is a vector

$$\alpha = (\alpha_1, \alpha_2, \dots, \alpha_n)$$

whose components are nonnegative integers. For any multi-index α, we define

$$|\alpha| = \alpha_1 + \alpha_2 + \cdots + \alpha_n; \tag{9.1}$$

moreover, for any vector $\mathbf{x} = (x_1, x_2, \ldots, x_n) \in \mathbb{R}^n$, we set

$$\mathbf{x}^\alpha = x_1^{\alpha_1} x_2^{\alpha_2} \cdots x_n^{\alpha_n}. \tag{9.2}$$

The following notation for partial derivatives is extremely convenient in writing PDEs:

$$D^\alpha = \frac{\partial^{|\alpha|}}{\partial x_1^{\alpha_1} \partial x_2^{\alpha_2} \cdots \partial x_n^{\alpha_n}}. \tag{9.3}$$

For example, if $\alpha = (1, 2)$, then

$$D^\alpha u = \frac{\partial^3 u}{\partial x_1 \partial x_2^2}. \tag{9.4}$$

We now consider a linear differential expression of the form

$$L(\mathbf{x}, D)u = \sum_{|\alpha| \leq m} a_\alpha(\mathbf{x}) D^\alpha u, \tag{9.5}$$

where $u : \mathbb{R}^n \to \mathbb{R}$. With this analytic operation on functions we associate an algebraic operation called the symbol.

Definition 1 *The* **symbol** *of the expression $L(\mathbf{x}, D)$ as given by (9.5) is*

$$L(\mathbf{x}, i\xi) := \sum_{|\alpha| \leq m} a_\alpha(\mathbf{x})(i\xi)^\alpha. \tag{9.6}$$

The **principal part** *of the symbol is*

$$L^p(\mathbf{x}, i\xi) := \sum_{|\alpha| = m} a_\alpha(\mathbf{x})(i\xi)^\alpha. \tag{9.7}$$

For instance, the symbol of Laplace's operator $\partial^2/\partial x_1^2 + \partial^2/\partial x_2^2$ is $-\xi_1^2 - \xi_2^2$. In an analogous fashion, we can associate a matrix-valued symbol with a system of PDEs. The definition of the principal part in this case is more involved; we shall address this issue later.

If the coefficients of the PDE are constant, and we are looking at solutions on all of space, the symbol is easily interpreted in terms of the Fourier transform: If

$$\hat{u}(\xi) := (2\pi)^{-n/2} \int_{\mathbb{R}^n} u(\mathbf{x}) \exp(-i\xi \cdot \mathbf{x})\, d\mathbf{x} \tag{9.8}$$

is the Fourier transform of $u(\mathbf{x})$, then $L(i\xi)\hat{u}(\xi)$ is the Fourier transform of $L(D)u(\mathbf{x})$.

In general, the symbol tells us how a differential expression acts on functions that have their support contained in a small neighborhood of a given point $\mathbf{x}$. If the coefficients are smooth, they are approximately constant in such a small neighborhood. Moreover, if u varies very rapidly, the highest-order derivatives are dominant over lower-order derivatives, and the principal part therefore contains the most important terms.

How the differential expression acts on rapidly varying functions of small support is of crucial importance for many basic properties of PDEs. The classification into types is based on the principal part of the symbol.

Let us first consider a second-order PDE in two space dimensions:

$$\begin{aligned} Lu &= a(x,y)u_{xx} + b(x,y)u_{xy} + c(x,y)u_{yy} \\ &\quad + d(x,y)u_x + e(x,y)u_y + f(x,y)u \\ &= g(x,y). \end{aligned} \tag{9.9}$$

The principal part of the symbol of L is

$$L^p(x,y;i\xi,i\eta) = -a(x,y)\xi^2 - b(x,y)\xi\eta - c(x,y)\eta^2. \tag{9.10}$$

Second-order PDEs are classified according to the behavior of L^p, viewed as a quadratic form in ξ and η. The quadratic form given by (9.10) can be represented in matrix form as

$$L^p(x,y;i\xi,i\eta) = (\xi,\eta)\begin{pmatrix} -a(x,y) & -\frac{1}{2}b(x,y) \\ -\frac{1}{2}b(x,y) & -c(x,y) \end{pmatrix}\begin{pmatrix} \xi \\ \eta \end{pmatrix}. \tag{9.11}$$

Definition 2 *The differential equation (9.9) is called* **elliptic** *if the quadratic form given by (9.10) is strictly definite,* **hyperbolic** *if it is indefinite, and* **parabolic** *if it is degenerate.*

The terms elliptic, parabolic, and hyperbolic are motivated by the analogy with the classification of conic sections.

Consider now a second-order PDE in n space dimensions:

$$Lu = a_{ij}(\mathbf{x})\frac{\partial^2 u}{\partial x_i \partial x_j} + b_i(\mathbf{x})\frac{\partial u}{\partial x_i} + c(\mathbf{x})u = 0. \tag{9.12}$$

Because the matrix of second partials of u is symmetric, we may assume without loss of generality that $a_{ij} = a_{ji}$. The principal symbol of this second-order PDE is still a quadratic form in ξ; we can represent this quadratic form as $\xi^T\mathbf{A}(\mathbf{x})\xi$, where $\mathbf{A}$ is the $n \times n$ matrix with components $-a_{ij}$.

Definition 3 *Equation (9.12) is called* **elliptic** *if all eigenvalues of* $\mathbf{A}$ *have the same sign,* **parabolic** *if* $\mathbf{A}$ *is singular, and* **hyperbolic** *if all but one of the eigenvalues of* $\mathbf{A}$ *have the same sign and one has the opposite sign. If* $\mathbf{A}$ *is nonsingular and there is more than one eigenvalue of each sign, the equation is called* **ultrahyperbolic**.

In this definition, it is understood that eigenvalues are counted according to their multiplicities.

The notion of characteristic surfaces is closely related to that of type. We make the following definition.

Definition 4 *The surface described by* $\phi(x_1, x_2, \ldots, x_n) = 0$ *is* **characteristic** *at the point* $\hat{\mathbf{x}}$ *if* $\phi(\hat{\mathbf{x}}) = 0$ *and, in addition,*

$$a_{ij}(\hat{\mathbf{x}})\frac{\partial \phi}{\partial x_i}(\hat{\mathbf{x}})\frac{\partial \phi}{\partial x_j}(\hat{\mathbf{x}}) = 0. \tag{9.13}$$

A surface is called characteristic if it is characteristic at each of its points.

In matrix form, condition (9.13) reads $(\nabla\phi)^T\mathbf{A}(\nabla\phi) = 0$. The matrix $\mathbf{A}$ is strictly definite, i.e., (9.12) is elliptic if and only if there are no nonzero real vectors with this property. We can therefore characterize elliptic equations as those without (real) characteristic surfaces.

For hyperbolic equations, on the other hand, all but one of the eigenvalues of $\mathbf{A}$ have the same sign, say, one eigenvalue is negative and the rest positive. Let $\mathbf{n}$ be a unit eigenvector corresponding to the negative eigenvalue. The span of $\mathbf{n}$ and its orthogonal complement are both invariant subspaces of $\mathbf{A}$, and, using the decomposition

$$\nabla\phi = (\mathbf{n}\cdot\nabla\phi)\mathbf{n} + (\nabla\phi - (\mathbf{n}\cdot\nabla\phi)\mathbf{n}), \tag{9.14}$$

we find

$$(\nabla\phi)^T\mathbf{A}(\nabla\phi) = -\lambda(\mathbf{n}\cdot\nabla\phi)^2 + [\nabla\phi - (\mathbf{n}\cdot\nabla\phi)\mathbf{n}]^T\mathbf{B}[\nabla\phi - (\mathbf{n}\cdot\nabla\phi)\mathbf{n}] = 0, \tag{9.15}$$

where $-\lambda$ is the negative eigenvalue of $\mathbf{A}$ and $\mathbf{B}$ is positive definite on the $(n-1)$-dimensional subspace perpendicular to $\mathbf{n}$. Let us now regard $\nabla\phi - (\mathbf{n}\cdot\nabla\phi)\mathbf{n}$, i.e., the part of $\nabla\phi$ that is perpendicular to $\mathbf{n}$, as given. Then $\mathbf{n}\cdot\nabla\phi$ can be determined from (9.15). For any nonzero choice of the perpendicular part of $\nabla\phi$, we get two real and distinct solutions for $\mathbf{n}\cdot\nabla\phi$.

Note that if we take any C^2 function $u:\mathbb{R}\to\mathbb{R}$ and compose it with ϕ, the resulting function satisfies

$$L^p u(\phi) = a_{ij}(\mathbf{x})\frac{\partial^2 u}{\partial x_i \partial x_j} = u''(\phi)\left[a_{ij}(\mathbf{x})\frac{\partial\phi}{\partial x_i}\frac{\partial\phi}{\partial x_j}\right] + u'(\phi)a_{ij}(\mathbf{x})\frac{\partial^2\phi}{\partial x_i\partial x_j}, \tag{9.16}$$

and if the surfaces $\phi = \text{const.}$ are characteristic, the coefficient of $u''(\phi)$ on the right-hand side vanishes. That is, the function $u(\phi)$ satisfies the equation $Lu = 0$ "to leading order." Because of this property, characteristics are important in the study of singularities of solutions of PDEs. Partial differential equations can have solutions that are (or whose derivatives are) discontinuous across a characteristic surface. For example, functions of the form $F(x-t) + G(x+t)$ satisfy the weak form of the wave equation $u_{tt} = u_{xx}$ even when F and G are discontinuous. The lines $x \pm t = \text{const.}$ are the characteristics of this equation.

The generalization of the definitions above to equations of higher order than second is straightforward.

Definition 5 *Let L be the mth-order operator defined in (9.5).* **Characteristic surfaces** *are defined by the equation*

$$L^p(\mathbf{x}, \nabla\phi) = 0. \tag{9.17}$$

An equation is called **elliptic** *at* $\mathbf{x}$ *if there are no real characteristics at* $\mathbf{x}$ *or, equivalently, if*

$$L^p(\mathbf{x}, i\xi) \neq 0 \quad \forall \xi \neq 0. \tag{9.18}$$

An equation is called **strictly hyperbolic** *in the direction* $\mathbf{n}$ *if*

1. $L^p(\mathbf{x}, i\mathbf{n}) \neq 0$ *and*

2. *all the roots* ω *of the equation*

$$L^p(\mathbf{x}, i\xi + i\omega\mathbf{n}) = 0 \tag{9.19}$$

are real and distinct for every $\xi \in \mathbb{R}^n$ *which is not collinear with* $\mathbf{n}$.

In applications, $\mathbf{n}$ is usually a coordinate direction associated with time. In this case, let us set $\mathbf{x} = (x_1, x_2, \ldots, x_{n-1}, t)$ and let $\xi = (\xi_1, \ldots, \xi_{n-1}, 0)$ be a spatial vector.

For rapidly oscillating functions of small support, we may think of the coefficients of L^p as approximately constant; let us assume they are constant. If ω is a root of (9.19), then $u = \exp(i(\xi \cdot \mathbf{x}) + i\omega t)$ is a solution of $L^p u = 0$. If ω has negative imaginary part, then this solution grows exponentially in time. Moreover, since L^p is homogeneous of degree m, i.e., $L^p(\mathbf{x}, \lambda(i\xi + i\omega\mathbf{n})) = \lambda^m L^p(\mathbf{x}, i\xi + i\omega\mathbf{n})$ for any scalar λ, there are always roots with negative imaginary parts if there are any roots which are not real (if we change the sign of ξ, we also change the sign of ω). Moreover, if we multiply ξ by a scalar factor λ, then ω is multiplied by the same factor, and hence solutions grow more and more rapidly the faster they oscillate in space. The condition that the roots in (9.19) are real is therefore a necessary condition for well-posedness of initial value problems.

We now turn our attention to systems of k PDEs involving k unknowns u_j, $j = 1, 2, \ldots, k$:

$$L_{ij}(\mathbf{x}, D)u_j = 0, \quad i = 1, 2, \ldots, k. \tag{9.20}$$

As for systems of algebraic equations, well-posed problems require equal numbers of equations and unknowns, so we shall assume that the operators L_{ij} form a square matrix $\mathbf{L}$. The generalization of the notions above is in principle quite straightforward.

Definition 6 **Characteristic surfaces** *are defined by the equation*

$$\det \mathbf{L}^p(\mathbf{x}, \nabla\phi) = 0, \tag{9.21}$$

and equations without real characteristic surfaces are called **elliptic**. **Strict hyperbolicity** *is also defined as in Definition 5, with* L^p *in the definition replaced by* $\det \mathbf{L}^p$. *A system in which all components of* $\mathbf{L}^p$ *are operators of first order is called* **evolutionary** *in the direction* $\mathbf{n}$ *if, for* ξ *not collinear with* $\mathbf{n}$, *all eigenvalues* ω *of the problem*

$$\det \mathbf{L}^p(\mathbf{x}, i\xi + i\omega\mathbf{n}) = 0$$

are real. The system is called **hyperbolic** *(not necessarily strictly) in the direction* $\mathbf{n}$ *if*

1. *det* $\mathbf{L}^p(\mathbf{x}, \mathbf{n}) \neq 0$ *and*

2. *for* ξ *not collinear with* $\mathbf{n}$, *all eigenvalues* ω *of the problem*

$$\det \mathbf{L}^p(\mathbf{x}, i\xi + i\omega\mathbf{n}) = 0$$

are real and there is a complete set of eigenvectors.

Note that, since we assumed that the components of $\mathbf{L}^p$ are of first order, we have $\mathbf{L}^p(\mathbf{x}, i\xi + i\omega\mathbf{n}) = \mathbf{L}^p(\mathbf{x}, i\xi) + \omega\mathbf{L}^p(\mathbf{x}, i\mathbf{n})$; hence the problem $\det \mathbf{L}^p = 0$ is a matrix eigenvalue problem for ω. If the eigenvalues are distinct, there is always a complete set of eigenvectors; hence strict hyperbolicity implies hyperbolicity.

In general, we need to be careful about defining the principal part of a system. A naive approach of simply taking the terms of highest order turns out to be unsatisfactory.

To see the problem, let us consider Laplace's equation in two dimensions,

$$u_{xx} + u_{yy} = 0, \tag{9.22}$$

and rewrite it as a system of first-order equations by setting $v = u_x$, $w = u_y$. The resulting system is

$$u_x = v, \quad u_y = w, \quad v_x + w_y = 0. \tag{9.23}$$

If we define $\mathbf{L}^p$ to be the part involving first-order terms, it is easy to see that $\det \mathbf{L}^p$ turns out to be identically zero. On the other hand, since Laplace's equation is the standard example of an elliptic equation, it would be desirable to have the equivalent first-order system also defined as elliptic. Obviously, we cannot then throw away the terms v and w in the first two equations of (9.23).

The difficulty is resolved by assigning "weights" s_i to each equation and t_j to each dependent variable in such a way that the order of each operator L_{ij} does not exceed $s_i + t_j$. The principal part L_{ij}^p is then defined to consist of those terms that have order exactly equal to $s_i + t_j$. We assume that the weights can be assigned in such a way that $\det \mathbf{L}^p$ does not vanish identically. In (9.23), for example, we would set $s_1 = s_2 = t_2 = t_3 = 0$ and $t_1 = s_3 = 1$. (Here it is understood that the ordering of the variables is u, v, w.) With these weights, the principal part of (9.23) is actually identical to (9.23), and we compute

$$\det \mathbf{L}^p(i\xi) = \det \begin{pmatrix} i\xi_1 & -1 & 0 \\ i\xi_2 & 0 & -1 \\ 0 & i\xi_1 & i\xi_2 \end{pmatrix} = -\xi_1^2 - \xi_2^2, \tag{9.24}$$

which is equal to the symbol of Laplace's equation.

For nonlinear equations and systems, the type can depend not only on the point in space but on the solution itself. We simply linearize the equation at a given solution and define the type to be that of the linearized equation. Characteristic surfaces are similarly defined as the characteristic surfaces of the linearized equation.

9.2 Application to viscoelastic flow

In this section, we shall briefly discuss how the notions introduced above apply to the equations governing viscoelastic flows. For the sake of concreteness, we shall limit our discussion to the Johnson–Segalman model. For a far more extensive discussion of change of type in viscoelastic flows, we refer to the literature, e.g., [40], [41].

The equations are

$$\begin{aligned} \rho\left(\frac{\partial \mathbf{v}}{\partial t} + (\mathbf{v}\cdot\nabla)\mathbf{v}\right) &= \operatorname{div}\mathbf{T} - \nabla p, \\ \operatorname{div}\mathbf{v} &= 0, \\ \frac{\partial \mathbf{T}}{\partial t} + (\mathbf{v}\cdot\nabla)\mathbf{T} - \frac{1+a}{2}(\nabla\mathbf{v}\mathbf{T} + \mathbf{T}(\nabla\mathbf{v})^T) & \\ +\frac{1-a}{2}(\mathbf{T}\nabla\mathbf{v} + (\nabla\mathbf{v})^T\mathbf{T}) + \lambda\mathbf{T} &= \mu(\nabla\mathbf{v} + (\nabla\mathbf{v})^T). \end{aligned} \tag{9.25}$$

This is a quasi-linear system of ten equations (six in 2-D flows), which we can put in the form

$$\mathbf{A}_0(\mathbf{q})\mathbf{q}_t + \mathbf{A}_1(\mathbf{q})\mathbf{q}_x + \mathbf{A}_2(\mathbf{q})\mathbf{q}_y + \mathbf{A}_3(\mathbf{q})\mathbf{q}_z = \mathbf{F}(\mathbf{q}). \tag{9.26}$$

To investigate characteristics, we need to consider the equation

$$\det\left(\omega\mathbf{A}_0 + \sum_{l=1}^{3}\xi_l\mathbf{A}_l\right) = 0. \tag{9.27}$$

This equation can be shown to reduce to the following:

$$\begin{aligned} &|\xi|^2\beta^4\left(\rho\beta^2 - |\xi|^2\left(\mu + \frac{1+a}{2}T_{aa} - \frac{1-a}{2}\Lambda_1\right)\right) \\ &\times\left(\rho\beta^2 - |\xi|^2\left(\mu + \frac{1+a}{2}T_{aa} - \frac{1-a}{2}\Lambda_2\right)\right) = 0. \end{aligned} \tag{9.28}$$

Here,

$$\beta = \omega + v_1\xi_1 + v_2\xi_2 + v_3\xi_3,\ T_{aa} = \mathbf{n}\cdot\mathbf{T}\mathbf{n}, \tag{9.29}$$

where $\mathbf{n}$ is a unit vector in the direction of (ξ_1, ξ_2, ξ_3). Moreover, with $\mathbf{P}$ denoting the orthogonal projection along $\mathbf{n}$, Λ_1 and Λ_2 are the eigenvalues of $\mathbf{PTP}$. In two dimensions, we get β^2 instead of β^4, and there is only one of the last two factors.

The following consequences can be deduced:

1. Well-posedness: If ω as determined by (9.28) is not real, localized short wave disturbances can be expected to grow in a catastrophic manner, and ill-posedness of the initial value problem occurs. It turns out that solutions of (9.28) with ω not real exist if one of the terms

$$\mu + \frac{1+a}{2}T_{aa} - \frac{1-a}{2}\Lambda_i \tag{9.30}$$

is negative. This is the case for an appropriate choice of ξ if

$$\frac{1-a}{2}\Lambda_{\text{max}} - \frac{1+a}{2}\Lambda_{\text{min}} > \mu, \tag{9.31}$$

where Λ_{max} and Λ_{min} are the largest and smallest eigenvalues of $\mathbf{T}$. Eigenvalues of $\mathbf{T}$ are greater than $-\mu/a$ if a is positive and less than $-\mu/a$ if a is negative. Hence no ill-posedness occurs if $a = \pm 1$.

2. Change of type in steady flow: This occurs if the fluid speed exceeds the speed of propagation of shear waves. For $\mathbf{T} = \mathbf{0}$, the condition becomes

$$\rho|v|^2 > \mu. \tag{9.32}$$

For nonzero $\mathbf{T}$, wave speeds become stress dependent and anisotropic. This change of type is analogous to transonic flow in gas dynamics. Below we shall discuss the phenomenon of delayed die swell, which is analogous to a gas dynamic shock.

3. Boundary conditions: For subcritical flows (i.e., the speed of the fluid is less than the wave speed), we need four conditions at an inflow boundary in addition to the usual Newtonian boundary conditions. If the flow becomes supercritical, we have to drop a velocity boundary condition at the outflow boundary, and we need an extra condition at the inflow boundary. For details, we refer to [79].

9.3 Melt fracture and spurt

In the extrusion of polymer melts from a pipe, an instability occurs which manifests itself in a sudden increase of flow rate and in distortions of the extrudate surface. In many polymer melts the instability starts as "sharkskin." This means that there are small-scale irregularities on the extrudate surface, without any apparent major effect on the gross characteristics of the flow. At higher flow rates, the extrudate tends to show distortions with a preferred period, which may or may not be axisymmetric. This often coincides with a stick-slip behavior, where the flow alternates between fast and slow. At yet higher flow rates, the stick-slip behavior disappears, and the extrudate becomes grossly distorted and wavy. Not all of these stages appear in all polymer melts. A review of the experimental results on melt fracture can be found in the article by Denn [6].

Since melt fracture is associated with an increase in the flow rate, it is often attributed to slip at the wall. An alternative explanation is a nonmonotone shear stress–shear rate response, where the observed spurt results from jumping to the higher value of the shear rate. Popular models allowing for such a response include the Johnson–Segalman, Giesekus, and Doi–Edwards models. While slip seems to be the preferred hypothesis among rheologists for melt fracture, there is some evidence for a nonmonotone constitutive law in wormlike micelles [100].

Over the past ten years, the dynamics of spurt in models with nonmonotone shear stress–shear rate curves, such as the Johnson–Segalman and Giesekus models (see Chapter 3), has been studied extensively [9, 46, 47, 51, 52, 53, 57]. These results concern the analysis of parallel shear flows with discontinuous shear rates, the approach to such flows from general initial data, and the possibility of oscillations. In all these works, it is assumed a priori that the flow is parallel shear flow. It was found [99], however, that the inclusion of two-dimensional disturbances leads to instabilities at the interface where the shear rate is discontinuous. These instabilities are analogous to coextrusion instabilities and are driven by a normal stress jump at the interface [98].

In this section, we shall not review the results on the dynamics of spurted flows, but limit ourselves to a discussion of how such flows relate to change of type. For generalized Newtonian fluids, a decreasing shear stress–shear rate function leads to ill-posedness [36]. This is not the case in viscoelastic flows. Indeed, the equations of a Johnson–Segalman fluid do not become ill posed in steady flow, even when the shear stress decreases with shear rate. Although instabilities occur, the growth rates as a function of wave number have an upper bound, and the dynamics remains well posed. There is, however, a change of type in the equations for one-dimensional steady flow.

We consider the Johnson–Segalman model with the addition of a Newtonian solvent viscosity. The governing equations are

$$\rho\left(\frac{\partial \mathbf{v}}{\partial t} + (\mathbf{v}\cdot\nabla)\mathbf{v}\right) = \operatorname{div}\mathbf{T} + \epsilon\Delta\mathbf{v} - \nabla p, \quad \operatorname{div}\mathbf{v} = 0, \tag{9.33}$$

where $\mathbf{T}$ is given by the constitutive equation

$$\frac{\partial \mathbf{T}}{\partial t} + (\mathbf{v}\cdot\nabla)\mathbf{T} - (\nabla\mathbf{v})\mathbf{T} - \mathbf{T}(\nabla\mathbf{v})^T + \lambda\mathbf{T} + \nu(\mathbf{TD} + \mathbf{DT}) = 2\mu\mathbf{D}. \tag{9.34}$$

(The relationship between ν and a in the previous section is $\nu = 1 - a$.) For steady parallel shear flows with a velocity field $\mathbf{v} = (v(y), 0, 0)$ and a pressure $p = Px + q(y)$, the equations become

$$\begin{aligned} T_{12}' + \epsilon v'' - P &= 0, \\ T_{22}' - q' &= 0, \\ (\nu - 2)v'T_{12} + \lambda T_{11} &= 0, \\ -v'T_{22} + \lambda T_{12} + \nu v'\frac{T_{11} + T_{22}}{2} &= \mu v', \\ \lambda T_{22} + \nu v' T_{12} &= 0. \end{aligned} \tag{9.35}$$

We can simplify the system further by introducing the combination

$$Z = -T_{22} + \frac{\nu}{2}(T_{11} + T_{22}). \tag{9.36}$$

We then find

$$\begin{aligned} T_{12}' + \epsilon v'' - P &= 0, \\ (Z - \mu)v' + \lambda T_{12} &= 0, \\ \lambda Z + (\nu^2 - 2\nu)v'T_{12} &= 0. \end{aligned} \tag{9.37}$$

The symbol associated with this system is

$$\begin{pmatrix} -\epsilon\xi^2 & i\xi & 0 \\ (Z-\mu)i\xi & \lambda & v' \\ T_{12}(\nu^2-2\nu)i\xi & (\nu^2-2\nu)v' & \lambda \end{pmatrix}. \tag{9.38}$$

On the other hand, we can solve the last two equations of (9.37) for T_{12} as a function of v'. The result is

$$T_{12} = \frac{\lambda\mu v'}{\lambda^2 + \nu(2-\nu)(v')^2}, \tag{9.39}$$

as already discussed in Chapter 3. The total shear stress at shear rate κ is then

$$\tau(\kappa) = \epsilon\kappa + \frac{\lambda\mu\kappa}{\lambda^2 + \nu(2-\nu)\kappa^2}. \tag{9.40}$$

It can be shown that the determinant of (9.38) is equal to

$$-(\lambda^2 + \nu(2-\nu)\kappa^2)\xi^2\tau'(\kappa), \tag{9.41}$$

i.e., a change of type occurs precisely when the shear stress becomes a decreasing function of shear rate.

Instability and/or ill-posedness can also result from the boundary conditions imposed on a PDE rather than from the equation itself. Conditions for loss of evolutionarity can be formulated at boundary points in a fashion quite analogous to what we did above for interior points. In the discussion above, we looked at the evolution of localized short wave disturbances. In doing so, we fixed the coefficients of the PDE at the constant values that they take at a particular point. We then considered solutions that are periodic in all spatial directions. If such solutions have a growth rate that tends to infinity as the spatial period tends to zero, then the problem is ill posed. In the neighborhood of a boundary point, we would instead look at a half-space problem and consider solutions of the PDE and boundary conditions which are periodic in directions tangent to the boundary but decay to zero exponentially in the direction normal to the boundary. Again we can ask the question whether such solutions grow in time.

The issue of well-posedness at the boundary arises in the context of allowing for slip. The simplest condition for slip is a Navier condition which says that the slip velocity is a function of the shear stress. If this is assumed, instabilities and possible ill-posedness arise precisely if the slip velocity decreases as a function of shear stress [63]. There are other possibilities, however. One possibility that has been considered is memory slip, where the slip velocity is a function of the shear stress history. It has been shown that such a hypothesis, in conjunction with the UCM constitutive model, can lead to instability and ill-posedness at the boundary [24, 78].

9.4 Delayed die swell

The phenomenon of die swell and delayed die swell was described in Chapter 1. A viscoelastic fluid emerging from a nozzle forms a jet whose diameter is substantially larger than that of the nozzle. Usually this swelling of the jet occurs immediately when the fluid leaves the nozzle, but in some cases it can be delayed. Joseph [41] was the first to link this phenomenon to change of type. The analysis presented in this section is due to Entov [8].

The analysis is based on a one-dimensional approximation where all quantities are regarded as homogeneous across the jet. If x is the coordinate along the axis of the jet, A the cross-sectional area, and v the (axial) velocity of the fluid, then conservation of mass requires that

$$A_t + (Av)_x = 0. \tag{9.42}$$

Moreover, with F denoting the force in the jet and ρ the density, the momentum balance reads

$$\rho((Av)_t + (Av^2)_x) = F_x. \tag{9.43}$$

The force is given by $(T_{xx} - p)A$, where $\mathbf{T}$ is the stress tensor. We assume that the constitutive law is that of the UCM fluid, and that $\mathbf{T}$ and $\nabla\mathbf{v}$ are approximately diagonal. Thus,

$$\nabla\mathbf{v} = \begin{pmatrix} v_x & 0 & 0 \\ 0 & -v_x/2 & 0 \\ 0 & 0 & -v_x/2 \end{pmatrix}. \tag{9.44}$$

Because of axisymmetry, $T_{yy} = T_{zz}$, and the constitutive law reduces to

$$\begin{aligned} (T_{xx})_t + v(T_{xx})_x - 2v_x T_{xx} + \lambda T_{xx} &= 2\mu v_x, \\ (T_{yy})_t + v(T_{yy})_x + v_x T_{yy} + \lambda T_{yy} &= -\mu v_x. \end{aligned} \tag{9.45}$$

Moreover, on the free surface of the jet, we have $T_{yy} = T_{zz} = p$, and since we assume homogeneous stresses across the jet, we have $F = A(T_{xx} - T_{yy})$. We are interested in steady states, so we set time derivatives equal to zero. To simplify notation, we also set $T_{xx} = \sigma$, $T_{yy} = \tau$. The system of equations now reads

$$\begin{aligned} (Av)' &= 0, \\ \rho(Av^2)' - (A(\sigma - \tau))' &= 0, \\ v\sigma' - 2v'\sigma + \lambda\sigma &= 2\mu v', \\ v\tau' + v'\tau + \lambda\tau &= -\mu v'. \end{aligned} \tag{9.46}$$

Inside the nozzle, we apply the same one-dimensional approximation (this is wrong, of course, but a reasonable assumption for a simple qualitative theory). The difference, however, is that the free surface condition does not apply and instead A is given. Hence, instead of (9.46) we have the equations

$$\begin{aligned} v' &= 0, \\ v\sigma' - 2v'\sigma + \lambda\sigma &= 2\mu v', \\ v\tau' + v'\tau + \lambda\tau &= -\mu v', \\ A &= A_0. \end{aligned} \tag{9.47}$$

To formulate shock conditions, we put the constitutive equations in conservation form:

$$\begin{aligned}\left(\frac{\sigma+\mu}{v^2}\right)' + \lambda\frac{\sigma}{v^3} &= 0,\\ (v(\tau+\mu))' + \lambda\tau &= 0.\end{aligned} \tag{9.48}$$

At the nozzle exit $x = 0$, we should thus have continuity of Av, $(\sigma+\mu)/v^2$, and $v(\tau+\mu)$. These values are therefore determined by the upstream flow in the nozzle:

$$A(0)v(0) = C_1,\ \frac{\sigma(0)+\mu}{v(0)^2} = C_2,\ v(0)(\tau(0)+\mu) = C_3. \tag{9.49}$$

In the case of regular die swell, we have a jet for $x > 0$, which is stress free, $\sigma = \tau$, and has uniform velocity $v' = 0$. We then satisfy (9.46) if

$$A(x) = A(0),\ v(x) = v(0),\ \sigma(x) = \tau(x) = \sigma(0)\exp(-\lambda x/v(0)). \tag{9.50}$$

Moreover, we satisfy the initial conditions (9.49) if

$$v(0)^3 = C_3/C_2,\ \sigma(0) = C_2 v(0)^2 - \mu,\ A(0) = C_1/v(0). \tag{9.51}$$

This solution is possible for all positive values of C_1, C_2, and C_3. Whether this solution is the physically relevant one, however, depends on the characteristic speeds of the hyperbolic system

$$\begin{aligned} A_t + (Av)_x &= 0,\\ \rho((Av)_t + (Av^2)_x) - (A(\sigma-\tau))_x &= 0,\\ \sigma_t + v\sigma_x - 2v_x\sigma + \lambda\sigma &= 2\mu v_x,\\ \tau_t + v\tau_x + v_x\tau + \lambda\tau &= -\mu v_x.\end{aligned} \tag{9.52}$$

(The characteristic speeds of the system $\mathbf{A}\mathbf{q}_t + \mathbf{B}\mathbf{q}_x = \mathbf{F}$ are the roots of the equation $\det(-c\mathbf{A}+\mathbf{B}) = 0$.) The characteristic speeds for the system above are $c = v$ (twice) and $c = v \pm w$, where

$$w = \sqrt{\frac{\sigma + 2\tau + 3\mu}{\rho}}. \tag{9.53}$$

If $v < w$ (the "subsonic" case), then three characteristic speeds are positive and one is negative. In this case, it is correct to prescribe three boundary conditions at the nozzle exit $x = 0$. If $v > w$ ("supersonic" case), however, a fourth boundary condition can be imposed at the nozzle exit. The obvious choice is $A(0) = A_0$, i.e., the cross section of the jet is the same as that of the nozzle. Die swell does not occur at the exit of the nozzle. However, it is now possible to have a shock farther down the jet where a transition from supersonic to subsonic flow occurs. This shock is the delayed die swell. The conditions across the shock are that (Av), $\rho Av^2 - A(\sigma-\tau)$, $(\sigma+\mu)/v^2$, and $(\tau+\mu)v$ are continuous.

Chapter 10

Jet Breakup

10.1 Governing equations

As we already mentioned in the introductory chapter, polymers greatly stabilize fluid jets and prevent or at least delay breakup into drops. In this chapter, we shall review some analytical results concerning this problem. The results discussed here are based primarily on [86] and [87]; we refer to the monograph of Yarin [105] for a review of earlier work on the problem.

The equations we shall consider are based on a one-dimensional approximation as in section 9.4 on delayed die swell. We shall, however, use a Lagrangian description. We consider a reference configuration in which the fluid jet is homogeneous, and we denote by X the position of a fluid particle in that reference configuration. The basic variable is the stretch $s(X,t)$, which is the factor by which the length of the jet (at the position occupied by particle X) has been stretched. If the radius of the jet in the reference configuration is δ, then the area of the cross section in the deformed configuration is $\pi\delta^2/s$ and the radius of the deformed jet is $\delta/\sqrt{s}$.

We shall assume creeping flow, i.e., we neglect the inertial terms in the momentum equation. However, we shall take surface tension forces into account. The force balance along the jet now becomes

$$F = \pi\delta^2 f(t) = \frac{\pi\delta^2}{s}(T_{xx} - p) + \frac{2\pi\sigma\delta}{\sqrt{s}}, \tag{10.1}$$

where σ is the surface tension coefficient and f depends only on t, but not on X. On the free surface of the jet, we have the normal stress balance

$$T_{yy} - p + \frac{\sigma\sqrt{s}}{\delta} = 0. \tag{10.2}$$

Combining these equations, we find

$$f(t) = \frac{T_{xx} - T_{yy}}{s} + \frac{\sigma}{\delta\sqrt{s}}. \tag{10.3}$$

The stresses T_{xx} and T_{yy} are determined by the constitutive equation. We note that the velocity gradient v_x is, in our current variables, given by s_t/s. We shall consider three cases:

1. The Newtonian fluid: In this case, we have $T_{xx} = 2\eta s_t/s$ and $T_{yy} = -\eta s_t/s$, where η is the viscosity.

2. The Oldroyd B fluid: In this case, we have

$$T_{xx} = 2\eta \frac{s_t}{s} + \Sigma, \; T_{yy} = -\eta \frac{s_t}{s} + \Upsilon, \tag{10.4}$$

   where, according to (9.45), Σ and Υ satisfy

$$\begin{aligned} \Sigma_t - 2\frac{s_t}{s}\Sigma + \lambda\Sigma &= 2\mu\frac{s_t}{s}, \\ \Upsilon_t + \frac{s_t}{s}\Upsilon + \lambda\Upsilon &= -\mu\frac{s_t}{s}. \end{aligned} \tag{10.5}$$

   We simplify this by the substitution $\Sigma = ps^2 - \mu$, $\Upsilon = qs^{-1} - \mu$, which yields the new equations

$$p_t + \lambda p = \lambda\mu s^{-2}, \; q_t + \lambda q = \lambda\mu s. \tag{10.6}$$

3. The Giesekus model: In this case, the left-hand side of (10.5) has the additional terms $\nu\Sigma^2$ and $\nu\Upsilon^2$, respectively.

We shall consider two different problems. One is the surface tension driven breakup of a jet. In that case, we consider periodic disturbances on an initially uniform jet. The average length of the jet does not change. This means that

$$\int_0^L s(X,t)\, dX = L, \tag{10.7}$$

where L is the period of the disturbance. The other situation is that of a filament in a stretching device. In this case, the length of the filament is a given, increasing function of time:

$$\int_0^L s(X,t)\, dX = \phi(t). \tag{10.8}$$

Here L is the length of the filament in the reference configuration. The constraint (10.7) or (10.8) implicitly determines the unknown force $f(t)$ in (10.3).

The model used in this section is capable of explaining some of the essential features of jet breakup, but it also has some weaknesses:

1. The one-dimensional approximation is valid as long as the fluid forms a slender jet, but it breaks down once drops begin to form.

2. We have neglected inertia, but, at least for the Newtonian case, it is known [17, 18] that inertial forces become important in the asymptotic approach to breakup.

10.2 Newtonian jet breakup

For the Newtonian fluid, we find the equation

$$3\eta s_t = f(t)s^2 - \frac{\sigma}{\delta}s^{3/2}. \tag{10.9}$$

Moreover, (10.7) implies that

$$f(t) = \frac{\sigma \int_0^L s^{3/2}(X,t)\,dX}{\delta \int_0^L s^2(X,t)\,dX}. \tag{10.10}$$

Whether the jet breaks in finite time (i.e., there is a time t_c and a point X_c such that $s(X_c,t) \to \infty$ as $t \to t_c$) depends on the initial data. For instance, if the initial condition $s(X,0)$ is such that the maximum value of s is assumed on a set of positive measure, then this will be so for all positive times, and consequently s cannot become infinite locally while maintaining a bounded integral. On the other hand, it was shown in [86] that there are initial data for which breakup in finite time does occur. The idea of the proof is simply to solve (10.9) backward in time, with an initial condition at the breakup time corresponding to unbounded s. For details of the argument, we refer to [86].

A combination of numerical computation and formal asymptotics is used in [87] to study the behavior close to breakup for a variety of initial data. The determining factor is how s behaves near the point where it has its maximum. For a typical smooth initial condition, we would have

$$s(X,0) \sim s_{max} - C(X - X_{max})^2. \tag{10.11}$$

If this is the case, then the maximum value of s behaves like $(t_c - t)^{-2}$, and consequently the radius of the jet is proportional to $t_c - t$ as the breakup time is approached. Papageorgiou [60, 61] has provided a more detailed asymptotic analysis which also determines the shape of the jet close to breakup. Like our analysis here, Papageorgiou's study is based on inertialess Stokes flow. On the other hand, Eggers [17, 18] has shown that inertial forces actually become important in the asymptotics of breakup. For highly viscous fluids, we expect the Stokes analysis to be valid for most of the evolution of the jet, but not all the way to breakup.

10.3 The Oldroyd B model

For the Oldroyd B fluid, we have the equations

$$\begin{aligned} 3\eta s_t + ps^3 - q + \frac{\sigma}{\delta}s^{3/2} &= f(t)s^2, \\ p_t + \lambda p &= \frac{\lambda\mu}{s^2}, \\ q_t + \lambda q &= \lambda\mu s \end{aligned} \tag{10.12}$$

and the constraint (10.7).

We shall prove that breakup in finite time does not occur and a continuous solution exists for all time. The only assumption on the initial data is that p, q, and s are continuous and positive. We define

$$m(t) = \min_{0\le X\le L} s(X,t), \quad M(t) = \max_{0\le X\le L} s(X,t). \tag{10.13}$$

From the second and third equations of (10.12), we can immediately conclude inequalities of the form

$$\begin{aligned} a(t) &\le p \le a(t) + b(t)\int_0^t m(\tau)^{-2}\,d\tau, \\ c(t) &\le q \le c(t) + d(t)\int_0^t M(\tau)\,d\tau, \end{aligned} \tag{10.14}$$

where a, b, c, d denote positive functions depending only on t and the initial data.

We now use the first equation of (10.12) at points where s has its maximum or its minimum. We obtain

$$\begin{aligned} 3\eta\dot{M} &\le -a(t)M^3 + c(t) + d(t)\int_0^t M(\tau)\,d\tau - \frac{\sigma}{\delta}M^{3/2} + f(t)M^2, \\ 3\eta\dot{m} &\ge -a(t)m^3 - b(t)m^3\int_0^t m(\tau)^{-2}\,d\tau + c(t)m - \frac{\sigma}{\delta}m^{3/2} + f(t)m^2. \end{aligned} \tag{10.15}$$

Here $\dot{M}$ and $\dot{m}$ denote right derivatives; in general M and m need not be differentiable.

Integrating the first equation of (10.12) over $[0, L]$, we find, moreover,

$$f(t) \ge -\int_0^L q\,dX \Big/ \int_0^L s^2\,dX. \tag{10.16}$$

Using the Cauchy–Schwarz inequality, we find

$$\int_0^L s^2\,dX \ge \left(\int_0^L s\,dX\right)^2 \Big/ L = L. \tag{10.17}$$

By integrating the third equation of (10.12), we easily obtain a bound on the integral of q, since the integral of s is equal to L. This yields a lower bound on f. By using this in the second inequality of (10.15), we find a lower bound on m.

Next, let

$$N(t) = \max_{0\le X\le L} ps - qs^{-2} + \frac{\sigma}{\delta}s^{-1/2}. \tag{10.18}$$

Let $\tilde{X}(t)$ be a point where the maximum is assumed. Then we must have $s_t(\tilde{X}(t),t) \le 0$, because otherwise the first equation of (10.12) would yield the conclusion that $s_t(X,t) > 0$ for every X, which is impossible according to (10.7). Next, we compute

$$\begin{aligned} \dot{N} &= p_t s - q_t s^{-2} + \left[p + 2qs^{-3} - \frac{1}{2}\frac{\sigma}{\delta}s^{-3/2}\right] s_t \\ &= -\lambda\left(N - \frac{\sigma}{\delta}s^{-1/2}\right) + \left[p + 2qs^{-3} - \frac{1}{2}\frac{\sigma}{\delta}s^{-3/2}\right] s_t. \end{aligned} \tag{10.19}$$

Here it is understood that the right-hand side is evaluated at the point $(\tilde{X}(t), t)$. If there is more than one point where the maximum in (10.18) is assumed, we have to take the point where the right-hand side in (10.19) is maximal. Now suppose N grows very large. Since we have an a priori bound for p, this is possible only if s becomes large. In that case, the expression in square brackets in (10.19) is positive. We conclude that $\dot{N}$ is negative if N is too large. Hence we obtain an upper bound on N. If $M(t)$ became infinite in finite time, then it follows easily from the definition (10.18) that $N(t)$ would also blow up. We have just seen that this is impossible, and hence solutions cannot blow up in finite time.

Instead of breaking up, jets of the Oldroyd B fluid evolve into a beads-on-a-string shape as shown in Figure 1.5. The drops remain connected by thin filaments whose radius thins exponentially with time [87].

For the Giesekus fluid, we do not have an analytical proof of breakup. However, numerical results show an initial evolution similar to the Oldroyd B case, but eventually breakup occurs with an asymptotic behavior similar to the Newtonian case.

10.4 Filament stretching

We shall now consider the case of filament stretching, i.e., the condition (10.8) with an increasing function $\phi(t)$ (typically ϕ is taken to be an exponential function). The question of interest is whether the stretching alone can cause breakup, without the effect of surface tension. Hence we now set $\sigma = 0$. In this case, perhaps surprisingly, the Newtonian fluid does not break up in finite time [32].

The governing equation is

$$\frac{s_t}{s^2} = f(t). \tag{10.20}$$

We have the constraint

$$\int_0^L s(X,t)\, dX = \phi(t). \tag{10.21}$$

We integrate (10.20) for each fixed value of X, resulting in

$$s(X,t) = \frac{1}{1/s(X,0) - \int_0^t f(\tau)\, d\tau}. \tag{10.22}$$

Assume now that there is a critical time t_0 and a point X_0 such that $s(X_0,t) \to \infty$ as $t \to t_0$. It follows from (10.22) that $s(X_0,0)$ must be the maximum of $s(X,0)$ and

$$\int_0^{t_0} f(\tau)\,d\tau = \frac{1}{s(X_0,0)}. \tag{10.23}$$

Suppose now that $s(X,0)$ is a smooth function. Then, for X near X_0, we have some $c > 0$ such that

$$\frac{1}{s(X,0)} \le \frac{1}{s(X_0,0)} + c(X - X_0)^2. \tag{10.24}$$

Suppose (10.24) holds for $X_0 \le X \le X_0 + \epsilon$. Then we find

$$\begin{aligned} \int_{X_0}^{X_0+\epsilon} s(X,t)\,dX &\ge \int_{X_0}^{X_0+\epsilon} \frac{1}{1/s(X_0,0) + c(X-X_0)^2 - \int_0^t f(\tau)\,d\tau}\,dX \\ &= \int_{X_0}^{X_0+\epsilon} \frac{1}{c(X-X_0)^2 + \int_t^{t_0} f(\tau)\,d\tau}\,dX. \end{aligned} \tag{10.25}$$

This tends to infinity as $t \to t_0$, contradicting the assumption that $\phi(t)$ stays finite. Indeed, if $\phi(t)$ is an exponential, then numerical simulations [32] show that the filament radius decreases exponentially with time.

It is interesting to contrast this behavior with the Giesekus model. For this model, McKinley (private communication) has undertaken numerical studies which suggest finite-time breakup. For the Giesekus model, we set $T_{11} = ps^2 - \mu$, $T_{yy} = q/s - \mu$, as we did above for the Oldroyd B. The equations resulting from this substitution are

$$\begin{aligned} p_t + \lambda p &= \frac{\mu\lambda}{s^2} - \nu\left(p^2 s^2 - 2p\mu + \frac{\mu^2}{s^2}\right), \\ q_t + \lambda q &= \mu\lambda s - \nu\left(\frac{q^2}{s} - 2q\mu + \mu^2 s\right), \\ ps^3 - q &= f(t)s^2. \end{aligned} \tag{10.26}$$

We now truncate these equations and retain only those terms which we expect to be dominant as breakup is approached. In this limit, s tends to infinity, and we also expect the stress T_{11} and hence ps^2 to be large. The dominant balance then leads to the truncated system

$$p_t = -\nu p^2 s^2, \; ps = f(t). \tag{10.27}$$

Combining the two equations, we find

$$p_t = -\nu f^2(t), \tag{10.28}$$

which can be integrated to yield

$$p(X,t) = p(X,t_0) - \nu \int_{t_0}^{t} f^2(s)\,ds. \tag{10.29}$$

Breakup occurs when p reaches zero, i.e., at a point X_c and time t_c where

$$p(X_c, t_c) = p(X_c, t_0) - \nu \int_{t_0}^{t_c} f^2(s)\, ds = 0. \tag{10.30}$$

We note that

$$p(X, t_c) - p(X_c, t_c) = p(X, t_0) - p(X_c, t_0) \sim C(X - X_c)^2 \tag{10.31}$$

for X near X_c. We have

$$s = \frac{f}{p}, \tag{10.32}$$

and near $t = t_c$, $X = X_c$, this is approximated by

$$s(t) \sim \frac{f(t)}{C(X - X_c)^2 + \nu \int_t^{t_c} f^2(s)\, ds}. \tag{10.33}$$

For the length to stay finite, it would then be necessary that

$$\frac{f(t)}{\sqrt{\nu \int_t^{t_c} f^2(s)\, ds}} \tag{10.34}$$

remain finite as $t \to t_c$. This is impossible. To see this, set

$$v(t) = \int_t^{t_c} f^2(s)\, ds. \tag{10.35}$$

We have $v(t) > 0$ and $\lim_{t \to t_c} v(t) = 0$. If

$$f(t) \leq K\sqrt{\nu v(t)}, \tag{10.36}$$

then

$$f^2(t) = -v'(t) \leq K^2 \nu v(t), \tag{10.37}$$

so

$$v'(t) + K^2 \nu v(t) \geq 0. \tag{10.38}$$

But this yields an exponential lower bound for $v(t)$, which is incompatible with $v(t_c) = 0$.

The question arises how we can reconcile this result with the apparent numerical observation of finite-time breakup. We first note from (10.28) that p_t is independent of X. Hence, locally near the point X_0 where p has its minimum, we can approximate p by

$$p(X, t) = C(X - X_0)^2 + q(t). \tag{10.39}$$

We have

$$q'(t) = -\nu f(t)^2. \tag{10.40}$$

Moreover,

$$s(X,t) = \frac{f(t)}{C(X-X_0)^2 + q(t)}, \tag{10.41}$$

which yields

$$\int_{X_0-\delta}^{X_0+\delta} s(X,t)\,dX = \frac{2f(t)\arctan(\sqrt{C/q(t)}\delta)}{\sqrt{Cq(t)}}. \tag{10.42}$$

For $q(t)$ small, we can approximate this by

$$\frac{\pi f(t)}{\sqrt{Cq(t)}}. \tag{10.43}$$

As breakup is approached, we expect this quantity to be close to $\phi(t)$, the prescribed length of the filament. We then find

$$f(t) = \frac{1}{\pi}\phi(t)\sqrt{Cq(t)}, \tag{10.44}$$

and consequently

$$q'(t) = -\frac{\nu}{\pi^2}Cq(t)\phi(t)^2. \tag{10.45}$$

If $\phi(t)$ is equal to $\exp(\gamma t)$, then $q(t)$ is proportional to

$$\exp\left(-\frac{\nu C}{2\gamma\pi^2}\exp(2\gamma t)\right), \tag{10.46}$$

which approaches zero extremely rapidly. Thus, although theoretically there is no finite-time breakup, there will be in practice, since the minimum thickness of the filament will quickly reach subatomic scales.

Bibliography

[1] A. Bahhar, J. Baranger, and D. Sandri, Galerkin discontinuous approximation of the transport equation and viscoelastic fluid flow on quadrilaterals, *Numer. Methods Partial Differential Equations* **14** (1998), 97–114.

[2] J. Baranger and A. Machmoum, Existence of approximate solutions and error bounds for viscoelastic fluid flow: characteristics method, *Comput. Methods Appl. Mech. Engrg.* **148** (1997), 39–52.

[3] J. Baranger and D. Sandri, Finite element approximation of viscoelastic fluid flow: existence of approximate solutions and error bounds. I. Discontinuous constraints, *Numer. Math.* **63** (1992), 13–27.

[4] B. Bernstein, E.A. Kearsley, and L.J. Zapas, A study of stress relaxation with finite strain, *Trans. Soc. Rheol.* **7** (1963), 391–410.

[5] R.B. Bird, C.F. Curtiss, R.C. Armstrong, and O. Hassager, *Dynamics of Polymeric Liquids*, 2nd ed., 2 vols., Wiley, New York 1987.

[6] M.M. Denn, Issues in viscoelastic fluid mechanics, *Ann. Rev. Fluid Mech.* **22** (1990), 13–34.

[7] P.G. Drazin and W.H. Reid, *Hydrodynamic Stability*, Cambridge University Press, Cambridge 1982.

[8] V.M. Entov, One-dimensional dynamics of jet and spinline flows of polymeric liquids, preprint.

[9] P. Español, X.F. Yuan, and R.C. Ball, Shear banding flow in the Johnson–Segalman fluid, *J. Non-Newt. Fluid Mech.* **65** (1996), 93–109.

[10] D.V. Boger and K. Walters, *Rheological Phenomena in Focus*, Elsevier, Amsterdam 1993.

[11] L. Boltzmann, Zur Theorie der elastischen Nachwirkung, *Ann. Phys. Chem.* **7** (1876), Ergänzungsband, 624–654.

[12] D. Brandon and W.J. Hrusa, Global existence of smooth shearing motions of a nonlinear viscoelastic fluid, *J. Integral Equations Appl.* **2** (1990), 333–351.

[13] M.J. Crochet, Numerical simulation of viscoelastic flow, in *Non-Newtonian Fluid Mechanics*, Von Karman Institute for Fluid Dynamics Lecture Series, Brussels, 1994-03, 1–68.

[14] B. Debbaut, On the inertial and extensional effects on the corner and lip vortices in a circular 4:1 abrupt contraction, *J. Non-Newt. Fluid Mech.* **37** (1990), 281–296.

[15] W.R. Dean and P.E. Montagnon, On the steady motion of viscous liquid in a corner, *Proc. Cambridge Phil. Soc.* **45** (1949), 389–394.

[16] C.D. Dimitropoulos, R. Sureshkumar, and A.N. Beris, Direct numerical simulation of viscoelastic turbulent channel flow exhibiting drag reduction: effect of the variation of rheological parameters, *J. Non-Newt. Fluid Mech.* **76** (1998), 433–468.

[17] J. Eggers and T.F. Dupont, Drop formation in a one-dimensional approximation to the Navier-Stokes equations, *J. Fluid Mech.* **262** (1994), 205–221.

[18] J. Eggers, Theory of drop formation, *Phys. Fluids* **7** (1995), 941–953.

[19] H. Engler, On the dynamic shear flow problem for viscoelastic liquids, *SIAM J. Math. Anal.* **18** (1987), 972–990.

[20] L. Gearhart, Spectral theory for contraction semigroups on Hilbert space, *Trans. Amer. Math. Soc.* **236** (1978), 385–394.

[21] H. Giesekus, A unified approach to a variety of constitutive models for polymer fluids based on the concept of configuration dependent molecular mobility, *Rheol. Acta* **21** (1982), 366–375.

[22] V. Girault and P.A. Raviart, *Finite Element Approximation of the Navier-Stokes Equations*, Springer Lecture Notes in Mathematics 749, Berlin 1979.

[23] M. Golubitsky, I. Stewart, and D.G. Schaeffer, *Singularities and Groups in Bifurcation Theory II*, Springer, New York 1988.

[24] M.D. Graham, Wall slip and the nonlinear dynamics of large amplitude oscillatory shear flows, *J. Rheol.* **39** (1995), 697–712.

[25] P. Grisvard, *Elliptic Problems in Nonsmooth Domains*, Pitman, Boston 1985.

[26] R. Guénette and M. Fortin, A new mixed finite element method for computing viscoelastic flows, *J. Non-Newt. Fluid Mech.* **60** (1995), 27–52.

[27] C. Guillopé and J.-C. Saut, Global existence and one-dimensional nonlinear stability of shearing motions of viscoelastic fluids of Oldroyd type, *RAIRO Modél. Math. Anal. Numér.* **24** (1990), 369–401.

[28] C. Guillopé and J.-C. Saut, Existence results for the flow of viscoelastic fluids with a differential constitutive law, *Nonlinear Anal.* **15** (1990), 849–869.

[29] C. Guillopé and J.-C. Saut, Existence and stability of steady flows of weakly viscoelastic fluids, *Proc. Roy. Soc. Edinburgh Sect. A* **119** (1991), 137–158.

[30] C. Guillopé and J.-C. Saut, Mathematical problems arising in differential models for viscoelastic fluids, in J.F. Rodrigues and A. Sequeira (eds.), *Mathematical Topics in Fluid Mechanics*, Longman, Harlow 1992, 64–92.

[31] T. Hagen and M. Renardy, Boundary layer analysis of the Phan-Thien–Tanner and Giesekus model in high Weissenberg number flow, *J. Non-Newt. Fluid Mech.* **73** (1997), 181–189.

[32] O. Hassager, M.I. Kolte, and M. Renardy, Failure and nonfailure of fluid filaments in extension, *J. Non-Newt. Fluid Mech.* **76** (1998), 137–151.

[33] I. Herbst, The spectrum of Hilbert space semigroups, *J. Operator Theory* **10** (1983), 87–94.

[34] E.J. Hinch, The flow of an Oldroyd fluid around a sharp corner, *J. Non-Newt. Fluid Mech.* **50** (1993), 161–171.

[35] F. Huang, Characteristic conditions for exponential stability of linear dynamical systems in Hilbert spaces, *Ann. Differential Equations* **1** (1985), 43–56.

[36] J. Hunter and M. Slemrod, Viscoelastic flow exhibiting hysteretic phase changes, *Phys. Fluids* **26** (1983), 2345–2351.

[37] D.F. James and K. Walters, A critical appraisal of available methods for the measurement of extensional properties of mobile systems, in A.A. Collyer (ed.), *Techniques in Rheological Measurement*, Chapman and Hall, London 1993, 33–53.

[38] M.W. Johnson and D. Segalman, A model for viscoelastic fluid behavior which allows non-affine deformation, *J. Non-Newt. Fluid Mech.* **2** (1977), 255–270.

[39] C. Johnson, U. Nävert, and J. Pitkäranta, Finite element methods for hyperbolic problems, *Comput. Methods Appl. Mech. Engrg.* **45** (1984), 285–312.

[40] D.D. Joseph, M. Renardy, and J.C. Saut, Hyperbolicity and change of type in the flow of viscoelastic fluids, *Arch. Rational Mech. Anal.* **87** (1985), 213–251.

[41] D.D. Joseph, *Fluid Dynamics of Viscoelastic Liquids*, Springer, New York 1990.

[42] A. Kaye, *Non-Newtonian flow in incompressible fluids*, College of Aeronautics, Cranfield, England, Tech. Note 134 (1962).

[43] J.U. Kim, Global smooth solutions for the equations of motion of a nonlinear fluid with fading memory, *Arch. Rational Mech. Anal.* **79** (1982), 97–130.

[44] R.C. King, M.R. Apelian, R.C. Armstrong, and R.A. Brown, Numerically stable finite element techniques for viscoelastic calculations in smooth and singular geometries, *J. Non-Newt. Fluid Mech.* **29** (1988), 147–216.

[45] H. Koch and D. Tataru, On the spectrum of hyperbolic semigroups, *Comm. Partial Differential Equations* **20** (1995), 901–937.

[46] R.W. Kolkka and G.R. Ierley, Spurt phenomena for the Giesekus viscoelastic fluid model, *J. Non-Newt. Fluid Mech.* **33** (1989), 305–323.

[47] R.W. Kolkka, D.S. Malkus, M.G. Hansen, G.R. Ierley, and R.A. Worthing, Spurt phenomena for the Johnson-Segalman and related models, *J. Non-Newt. Fluid Mech.* **29** (1988), 303–335.

[48] O. Ladyzhenskaya, *The Mathematical Theory of Viscous Incompressible Flow*, Gordon and Breach, New York 1963.

[49] R.G. Larson, Instabilities in viscoelastic flows, *Rheol. Acta* **31** (1992), 213–263.

[50] M. Laso and H.C. Oettinger, Calculation of viscoelastic flow using molecular models: the CONNFFESSIT approach, *J. Non-Newt. Fluid Mech.* **47** (1993), 1–20.

[51] D.S. Malkus, J.A. Nohel, and B.J. Plohr, Analysis of new phenomena in shear flow of non-Newtonian fluids, *SIAM J. Appl. Math.* **51** (1991), 899–929.

[52] D.S. Malkus, J.A. Nohel, and B.J. Plohr, Dynamics of shear flow of a non-Newtonian fluid, *J. Comput. Phys.* **87** (1990), 464–487.

[53] D.S. Malkus, J.A. Nohel, and B.J. Plohr, Oscillations in piston-driven shear flow of a non-Newtonian fluid, in J.F. Dijksman and G.D.C. Kuiken (eds.), *IUTAM Symposium on Numerical Simulation of Non-Isothermal Flow of Viscoelastic Liquids*, Kluwer Academic Publishers, Dordrecht, Boston 1995 57–71.

[54] J.M. Marchal and M.J. Crochet, A new mixed finite element for calculating viscoelastic flow, *J. Non-Newt. Fluid Mech.* **26** (1987), 77–114.

[55] S. Matušů-Nečasová, A. Sequeira, and J.H. Videman, Existence of classical solutions for compressible viscoelastic fluids of Oldroyd type past an obstacle, *Math. Methods Appl. Sci.* **22** (1999), 449–460.

[56] J.C. Maxwell, On the dynamical theory of gases, *Philos. Trans. Roy. Soc. London Ser. A* **157** (1867), 49–88.

[57] J.A. Nohel, R.L. Pego, and A.E. Tzavaras, Stability of discontinuous steady states in shearing motions of a non-Newtonian fluid, *Proc. Roy. Soc. Edinburgh* **115A** (1990), 39–59.

[58] J.G. Oldroyd, Non-Newtonian effects in steady motion of some idealized elastico-viscous liquids, *Proc. Roy. Soc. London Ser. A* **245** (1958), 278–297.

[59] P.J. Oliveira, F.T. Pinho, and G.A. Pinto, Numerical simulation of non-linear elastic flows with a general collocated finite-volume method, *J. Non-Newt. Fluid Mech.* **79** (1998), 1–43.

[60] D.T. Papageorgiou, Analytical description of the breakup of liquid jets, *J. Fluid Mech.* **301** (1995), 109–132.

[61] D.T. Papageorgiou, On the breakup of viscous liquid threads, *Phys. Fluids* **7** (1995), 1529–1544.

[62] A. Pazy, *Semigroups of Linear Operators and Applications to Partial Differential Equations*, Springer, New York 1983.

[63] J.R.A. Pearson and C.J.S. Petrie, On melt flow instability of extruded polymers, in R.E. Wetton and R.H. Whorlow (eds.), *Polymer Systems: Deformation and Flow*, Macmillan, London 1968, 163–187.

[64] N. Phan-Thien and R.I. Tanner, A new constitutive equation derived from network theory, *J. Non-Newt. Fluid Mech.* **2** (1977), 353–365.

[65] J. Prüß, On the generation of C_0 semigroups, *Trans. Amer. Math. Soc.* **284** (1984), 847–857.

[66] D. Rajagopalan, R.C. Armstrong, and R.A. Brown, Finite element methods for calculation of steady, viscoelastic flow using constitutive equations with a Newtonian viscosity, *J. Non-Newt. Fluid Mech.* **36** (1990), 159–192.

[67] M. Reiner, The Deborah number, *Physics Today* **17** (1) (1964), 62.

[68] M. Renardy, W.J. Hrusa, and J.A. Nohel, *Mathematical Problems in Viscoelasticity*, Longman, Harlow 1987.

[69] M. Renardy, Local existence theorems for the first and second initial-boundary value problem for a weakly non-Newtonian fluid, *Arch. Rational Mech. Anal.* **83** (1983), 229–244.

[70] M. Renardy, A local existence and uniqueness theorem for a K-BKZ fluid, *Arch. Rational Mech. Anal.* **88** (1985), 83–94.

[71] M. Renardy, Existence of slow steady flows of viscoelastic fluids with differential constitutive equations, *Z. Angew. Math. Mech.* **65** (1985), 449–451.

[72] M. Renardy, Inflow boundary conditions for steady flows of viscoelastic fluids with differential constitutive laws, *Rocky Mountain J. Math.* **18** (1988), 445–453 (Corrigendum **19** (1989), 561).

[73] M. Renardy, Existence of slow steady flows of viscoelastic fluids of integral type, *Z. Angew. Math. Mech.* **68** (1988), T40–T44.

[74] M. Renardy, Recent advances in the mathematical theory of steady flows of viscoelastic fluids, *J. Non-Newt. Fluid Mech.* **29** (1988), 11–24.

[75] M. Renardy, Existence of steady flows of Jeffreys fluids with traction boundary conditions, *Differential Integral Equations* **2** (1989), 431–437.

[76] M. Renardy, Local existence of solutions of the Dirichlet initial-boundary value problem for incompressible hypoelastic materials, *SIAM J. Math. Anal.* **21** (1990), 1369–1385.

[77] M. Renardy, An alternative approach to inflow boundary conditions for Maxwell fluids in three space dimensions, *J. Non-Newt. Fluid Mech.* **36** (1990), 419–425.

[78] M. Renardy, Short wave instabilities resulting from memory slip, *J. Non-Newt. Fluid Mech.* **35** (1990), 73–76.

[79] M. Renardy, A well-posed boundary value problem for supercritical flow of viscoelastic fluids of Maxwell type, in B.L. Keyfitz and M. Shearer (eds.), *Nonlinear Evolution Equations that Change Type*, IMA Vol. Math. Appl., 27, Springer, New York 1990, 181–191.

[80] M. Renardy, An existence theorem for model equations resulting from kinetic theories of polymer solutions, *SIAM J. Math. Anal.* **22** (1991), 313–327.

[81] M. Renardy, Compatibility conditions at corners between walls and inflow boundaries for fluids of Maxwell type, *Z. Angew. Math. Mech.* **71** (1991), 37–45.

[82] M. Renardy, A centre manifold theorem for hyperbolic PDEs, *Proc. Roy. Soc. Edinburgh* **122A** (1992), 363–377.

[83] M. Renardy, The stresses of an upper convected Maxwell fluid in a Newtonian velocity field near a reentrant corner, *J. Non-Newt. Fluid Mech.* **50** (1993), 127–134.

[84] M. Renardy, How to integrate the upper convected Maxwell (UCM) stresses near a singularity (and maybe elsewhere, too), *J. Non-Newt. Fluid Mech.* **52** (1994), 91–95.

[85] M. Renardy, On the linear stability of hyperbolic PDEs and viscoelastic flows, *Z. Angew. Math. Phys.* **45** (1994), 854–865.

[86] M. Renardy, Some comments on the surface-tension driven break-up (or the lack of it) of viscoelastic jets, *J. Non-Newt. Fluid Mech.* **51** (1994), 97–107.

[87] M. Renardy, A numerical study of the asymptotic evolution and breakup of Newtonian and viscoelastic jets, *J. Non-Newt. Fluid Mech.* **59** (1995), 267–282.

[88] M. Renardy, Existence of steady flows for Maxwell fluids with traction boundary conditions on open boundaries, *Z. Angew. Math. Mech.* **75** (1995), 153–155.

[89] M. Renardy, A matched solution for corner flow of the upper convected Maxwell fluid, *J. Non-Newt. Fluid Mech.* **58** (1995), 83–89.

[90] M. Renardy, Nonlinear stability of Jeffreys fluids at low Weissenberg numbers, *Arch. Rational Mech. Anal.* **132** (1995), 37–48.

[91] M. Renardy, Initial-value problems with inflow boundaries for Maxwell fluids, *SIAM J. Math. Anal.* **27** (1996), 914–931.

[92] M. Renardy, The high Weissenberg number limit of the UCM model and the Euler equations, *J. Non-Newt. Fluid Mech.* **69** (1997), 293–301.

[93] M. Renardy, High Weissenberg number boundary layers for the upper convected Maxwell fluid, *J. Non-Newt. Fluid Mech.* **67** (1997), 125–132.

[94] M. Renardy, Re-entrant corner behavior of the PTT fluid, *J. Non-Newt. Fluid Mech.* **69** (1997), 99–104.

[95] M. Renardy, Asymptotic structure of the stress field in flow past a cylinder at high Weissenberg number, *J. Non-Newt. Fluid Mech.* **90** (2000), 13–23.

[96] M. Renardy, Wall boundary layers for Maxwell liquids, *Arch. Rational Mech. Anal.*, to appear.

[97] M. Renardy and R.C. Rogers, *An Introduction to Partial Differential Equations*, Springer, New York 1993.

[98] Y. Renardy, Stability of the interface in two-layer Couette flow of upper convected Maxwell liquids, *J. Non-Newt. Fluid Mech.* **28** (1988), 99–115.

[99] Y. Renardy, Spurt and instability in a two-layer Johnson–Segalman fluid, *Theoret. Comput. Fluid Dyn.* **56** (1995), 463–475.

[100] N.A. Spenley, M.E. Cates, and T.C.B. McLeish, Nonlinear rheology of wormlike micelles, *Phys. Rev. Lett.* **71** (1993), 939–942.

[101] R. Temam, *Navier-Stokes Equations*, 3rd ed., North Holland, Amsterdam 1984.

[102] R. Temam, *Navier–Stokes Equations and Nonlinear Functional Analysis*, 2nd ed., SIAM, Philadelphia 1995.

[103] C.A. Truesdell and W. Noll, The nonlinear field theories of mechanics, in S. Flügge (ed.), *Handbuch der Physik III/3*, Springer, New York 1965.

[104] G.M. Wilson and B. Khomami, An experimental investigation of interfacial instabilities in multilayer flow of viscoelastic fluids. II. Elastic and nonlinear effects in incompatible polymer systems, *J. Rheol.* **37** (1993), 315–339.

[105] A.L. Yarin, *Free Liquid Jets and Films: Hydrodynamics and Rheology*, Longman, Harlow 1993.

Index